PERGAMON INTERNATIONAL LIBRA
of Science, Technology, Engineering and Social
The 1000-volume original paperback library in aid of edu
industrial training and the enjoyment of leisure
Publisher: Robert Maxwell, M.C.

INTRODUCTION TO MICROELECTRONICS

SECOND EDITION

INTRODUCTION TO MICROELECTRONICS

SECOND EDITION

BY

D. RODDY

M.Sc., C.Eng., M.I.E.E., M.I.E.R.E.

Professor of Electrical Engineering
Lakehead University, Ontario, Canada.

PERGAMON PRESS

OXFORD . TORONTO . SYDNEY

PARIS . NEW YORK . FRANKFURT

U.K.	Pergamon Press Ltd., Headington Hill Hall, Oxford OX3 0BW, England
U.S.A.	Pergamon Press Inc., Maxwell House, Fairview Park, Elmsford, New York 10523, U.S.A.
CANADA	Pergamon of Canada Ltd., 75 The East Mall, Toronto, Ontario, Canada
AUSTRALIA	Pergamon Press (Aust.) Pty. Ltd., 19a Boundary Street, Rushcutters Bay, N.S.W. 2011, Australia
FRANCE	Pergamon Press SARL, 24 rue des Ecoles, 75240 Paris, Cedex 05, France
FEDERAL REPUBLIC OF GERMANY	Pergamon Press GmbH, 6242 Kronberg Taunus, Pferdstrasse 1, Frankfurt, Federal Republic of Germany

First edition 1970

Second edition 1978

British Library Cataloging in Publication Data

Roddy, Dennis
Introduction to microelectronics.—2nd ed.—(Pergamon international library).
1. Microelectronics
I. Title
621.381'73 TK7874 77-30485

ISBN 0-08-022687-6 (Hard cover)
ISBN 0-08-022688-4 (Flexicover)

Printed in Great Britain at the University Press, Aberdeen.

Contents

4. Bipolar Logic Circuits

5. Differential Amplifiers

6. Metal–Insulator–Semiconductor Devices

7. MOS Circuits

Preface to Second Edition

SINCE the appearance of the first edition in 1970, very rapid progress has been, and is being, made in microelectronics, especially in the area of semiconductor memories. Large scale integrated (LSI) circuits are now commonplace, and wide new areas of application are to be found in calculators, mini-computers, and microprocessors, the latter having a profound effect in the fields of communications, control, instrumentation, and business machines.

Whereas at one time it was thought that bipolar technology would provide high speed, medium density circuits at moderate cost, and MOS technology low speed, high density circuits at low cost, in fact, developments in the two technologies appear to be "leap-frogging" so that high speed, high density, circuits at comparable costs are available in both.

A new chapter on semiconductor memories has been included in this edition. Some of the existing chapters have been expanded to include details of the developments which appear to be the most important, and the introductory basic theory in Chapter 1 has been shortened somewhat. The changes have meant some reorganization of the book, but it follows essentially the plan of the first edition. The opportunity has been taken to correct those errors which have come to light, and to clarify the text in places.

As for the first edition, I am greatly indebted to many sources for information, in addition to those acknowledged in the text. I would like particularly to thank L. J. M. Esser of Philips Research Laboratories, Eindhoven, for providing very comprehensive details of charge coupled devices. Also H. H. Berger and S. K. Wiedmann of IBM, Component Development, Boeblingen, for the very comprehensive details of merged transistor logic; to J. E. Coe and W. Oldham of Intel. Corp. for details of the Intel. 16k RAM. Again, I must point out that much of the data and information supplied from all sources has had to be compressed for a book of this size, and sole responsibility rests with me for the accuracy of presentation.

D. RODDY

Thunder Bay, Ontario
June 1977

Preface to Second Edition

Preface to First Edition

SINCE the introduction of the transistor, the electronic properties of solids in general, and semiconductors in particular, have been exploited at a remarkable rate. New devices are being reported on almost continuously in the technical literature; some utilizing optical–electronic coupling effects, others acoustic–electronic couplings, yet others utilizing cold-cathode, hot-electron mission. In a book of this nature it is not possible to keep abreast of all these potentially useful devices, and only those devices and circuits which are reasonably well established in practical applications are described.

Perhaps the most startling advantage which accompanies the new technology is the reduction in physical size of components and circuits, but other, no less important advantages accrue, such as increased reliability, reduction in cost for complex circuit functions, and reduction of power consumption. Also, very complex circuit functions become practical possibilities rather than just remaining interesting theoretical solutions to engineering problems. There are of course practical limitations to the size of circuit that can be accepted in microelectronics, the most serious of these being power dissipation, and the cost of testing completed assemblies. Thus, although the power consumption per circuit function may be reduced, the very high component packing density in integrated circuits means that the power density (watts per unit area) can be very high. Secondly, the most costly operation in making integrated circuits lies in the encapsulation and packaging of the circuits, and therefore circuit functions must be tested prior to this; the more complex the circuit, the less versatile it is, and if high reliability is to be assured, testing will be costly. Because of these limitations it would seem therefore that large-scale integration (LSI) and medium-scale integration (MSI) will evolve gradually from the presently available integrated circuits. A good critical review of LSI is presented in *Electronics*, vol. 41, No. 12.

It is hoped that this book will be useful for the older engineer who has become familiar with discrete transistors and who is now faced with the problem of becoming familiar with integrated circuit technology and the world of microelectronics; also to the younger technologist who wishes to know more about these topics.

Although specific acknowledgements are made in the text for permission to reprint diagrams and other information, such acknowledgements do not do full justice to the many companies who provided handbooks, textbooks, and technical literature all of which was used freely as sources of information and for which the author is indeed grateful. It was kindly pointed out by the Plessey company that in this rapidly advancing field of technology problems exist regarding accuracy of literature and this can become obsolete very quickly with advances and improvements in products and process development.

Grateful acknowledgement is also made to Mr. Ian Thomson, who provided a very comprehensive write-up on the Silicon Avalanche Diode; also to the professional institutions and to the publishers of various technical journals for their kind co-operation in supplying information. It should be pointed out, however, that much of the information and data has had to be compressed for a book of this size, and sole responsibility rests with the author for the accuracy of its presentation.

D. RODDY

Port Arthur, Ontario

CHAPTER 1

Basic Theory

1.1. Introduction

Electrons in solids possess energy, the kinetic energy component arising through various motional states, and the potential energy mainly from the electrostatic field of the atomic nucleii. The total energy is quantized, that is, it can only be altered in discrete amounts (for example by application of an external voltage, or by incident radiation). In a solid, these discrete energy levels are grouped into bands as a result of interaction between the atomic nucleii, and the density of levels within a given band is sufficiently high for the energy range to be considered continuous (i.e. its discrete nature may be ignored within a band). The bands themselves may be separated by comparatively large energy gaps, in which no permissible energy levels occur, these being referred to as forbidden energy gaps.

Operation of solid state electronic devices and circuits results from the fact that energy levels within a solid can be shifted in a controlled manner, this giving rise to some remarkable properties not explainable on a simple electronic charge picture of electronic conduction.

1.2. Energy Bands in Solids

For electrical conduction to occur in a solid it must be possible for electrons within it to change their energetic states, since an externally applied voltage is going to impart energy to the electrons. This means that not only must there be electrons available for conduction, but "empty" energy levels must be available for these. Figure 1.1 shows a one-dimensional energy band diagram, which is simply an energy scale showing permissible and forbidden bands. The lower energy bands apply to electrons close to the nucleii, and as these bands are normally completely filled (the closed shells in chemical binding), they cannot contribute to electrical conduction. Only the two outermost bands, the valence band and the conduction band, are of importance in the operation of electronic devices. E_c denotes the lower edge of the conduction band, E_v the upper edge of the valence band, and the forbidden gap is $E_g = E_c - E_v$.

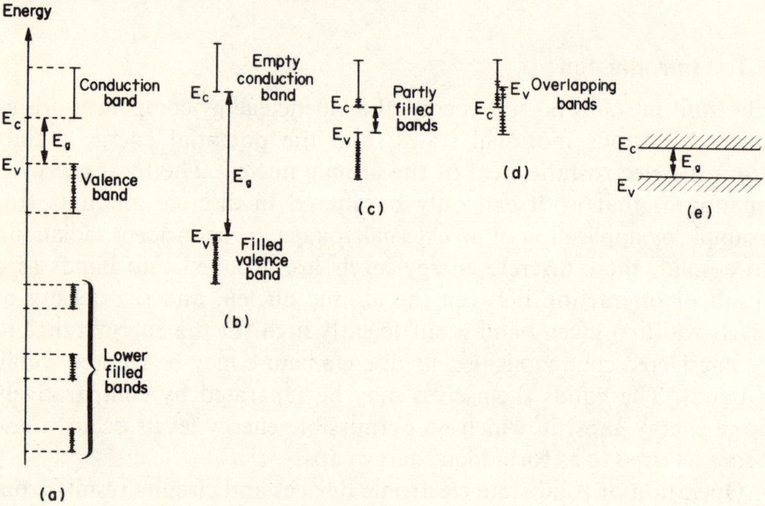

FIG. 1.1. Energy-band levels in solids: (a) general scheme; (b) insulator; (c) semiconductor; (d) metal; (e) extension to show dimension in crystal.

Figure 1.1b shows the situation for a divalent material, that is, one which has two valence electrons per atom. Here, all the energy states in the valence band are normally filled and providing the forbidden gap E_g is sufficiently large to prevent electrons being excited into the higher permissible levels in the conduction band, the material is an electrical insulator. Diamond, for example, has an energy band gap of 6 eV, which compared to the average thermal energy available at room temperature of about 1/40 eV, makes diamond a perfect insulator. (The electron volt (eV) is a unit of energy particularly appropriate for dealing with electron energies. It is equivalent to 1.6×10^{-19}

Joules, or 1.18×10^{19} ft-lbs, the latter being introduced simply to emphasize that it is an *energy* unit.)

Silicon has the same basic structure as diamond, but its band gap is 1.15 eV, and at room temperature some electrons will be excited from the valence band into the conduction band. The density of conduction band electrons at room temperature has been calculated as 1.4×10^{10} cm^{-3}, and of course there will be a corresponding density of empty energy states in the valence band. Both the conduction band and the valence band meet the conditions required for electrical conduction, and both will contribute to the conduction process. The energy band diagram is as shown in Fig. 1.1c. This is typical of a semiconductor for which the density of electrons in the conduction band is many orders of magnitude less than that for good metallic conductors, and for example, at low temperatures, silicon becomes a good insulator.

In dealing with the conduction process in the valence band it is vastly more convenient to treat the vacant energy levels as though they represented positive mass, positive charge particles analogous to electrons (which of course are treated as positive mass, negative charge particles). This approach can be justified on rigorous theoretical and experimental grounds. The fictitious positive mass, positive charge particles are termed *holes*, and electrons and holes referred to generally as *carriers* (meaning charge carriers). The electron and hole densities can be controlled independently of one another by the addition of suitable impurities to a semiconductor, giving rise to what are known as *extrinsic* semiconductors. These are discussed further in § 1.4. Where the effects of impurities are absent or can be ignored, the material is referred to as *intrinsic*.

Returning to the energy band diagram, Fig. 1.1d shows the situation for metals. Here, conduction and valence bands overlap providing an abundance of electrons, along with permissible energy states in the overlap region. Some of the divalent metals, for example zinc, exhibit predominantly hole type conduction. Gold, silver, and copper are monovalent metals—that is there is one valence electron per atom, and therefore the valence bands are not completely filled. This occurs in addition to band overlap, so that these metals are very good conductors, the conduction being predominantly by electron carriers.

It is usually assumed that energy bands extend throughout a material and the energy band diagram is modified as shown in Fig. 1.1e to show distance through material along the abscissa.

1.3. The Fermi Energy Level

The Fermi energy level is a sort of datum level of special significance in the theory of solids. Mathematically it may be defined as that level which has exactly a 50–50 chance of being occupied by an electron disregarding any restrictions imposed by band structure. For any structure in thermal equilibrium (i.e. no net transfer of carriers from one region to another), the Fermi level is the same throughout. This is true whether the structure is a single solid, or solids in contact, and the rule that *in thermal equilibrium the Fermi level has the same value throughout a structure*, provides a very simple and powerful way of illustrating how many semiconductor devices work. The other levels maintain their positions relative to the Fermi level.

As already mentioned, for an intrinsic semiconductor, the electron density in the conduction band is equal to the hole density in the valence band, and this density is known as the intrinsic carrier density n_i. The Fermi level for most intrinsic semiconductors falls about midway in the band gap, and so the energy-band diagram is as shown in Fig. 1.2a. Here the Fermi level is denoted by E_i (the *i* signifying *i*ntrinsic), and it is the shift of E_F from the E_i position that occurs when certain impurities are added that is of fundamental importance in semiconductor devices. This is discussed later.

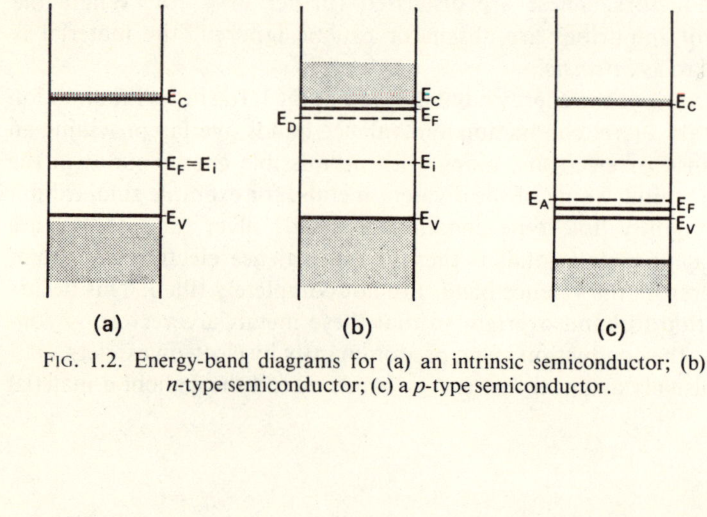

FIG. 1.2. Energy-band diagrams for (a) an intrinsic semiconductor; (b) an *n*-type semiconductor; (c) a *p*-type semiconductor.

1.4. Extrinsic Semiconductors

Extrinsic semiconductors are formed by adding impurities to an intrinsic semiconductor in such a way that the electron or hole density is increased (usually by many orders of magnitude). Impurities which contribute conduction electrons are termed *donor* impurities. These give rise to *n*-type extrinsic semiconductors (*n* for *n*egative). For example, the intrinsic semiconductor silicon has four valence electrons per atom, and by adding phosphorus which has five valence electrons per atom, an *n*-type semiconductor results. The donor impurity is positively ionized when it loses an electron but the donor itself does not contribute a hole to conduction because the ionized energy level is close to the conduction band and well removed from the valence band and, the impurity centres themselves are not free to move. The donor level is denoted by E_D as shown in Fig. 1.2b. The contribution of electrons due to donors may be a thousand or more times greater than the intrinsic density at room temperature, so that the Fermi level is shifted well up towards the conduction band, as shown in Fig. 1.2b. Conduction due to extrinsic electron density will be very much greater than the intrinsic conductivity.

A *p*-type impurity is one which captures an electron from the valence band and therefore effectively contributes a hole to the conduction process. Such impurities are termed *acceptor* impurities. For example, boron which has three valence electrons per atom is an acceptor type impurity in silicon. Since the acceptor impurity gains an electron it becomes negatively ionized. It does not contribute this electron to conduction because its ionized energy level is close to the valence band and well removed from the conduction band and the impurity centre is not free to move. The acceptor level is denoted by E_A, Fig. 1.2c.

With acceptor type impurities the Fermi level is shifted well down towards the valence band and for normal *p*-type doping the conductivity due to extrinsic hole density will be very much greater than the intrinsic conductivity.

When the temperature of an extrinsic semiconductor is increased well above room temperature, intrinsic conditions tend to return, that is the Fermi level tends to move towards the centre of the band-gap again. This is because electron–hole pairs are thermally generated (these making up the intrinsic carrier densities of holes and electrons), and at high temperatures the intrinsic carriers dominate, i.e. the electron density in the conduction band is about equal to the hole density in the valence band. The shift of the Fermi level with temperature is shown in Fig. 1.3 for both *p*- and *n*-type materials.

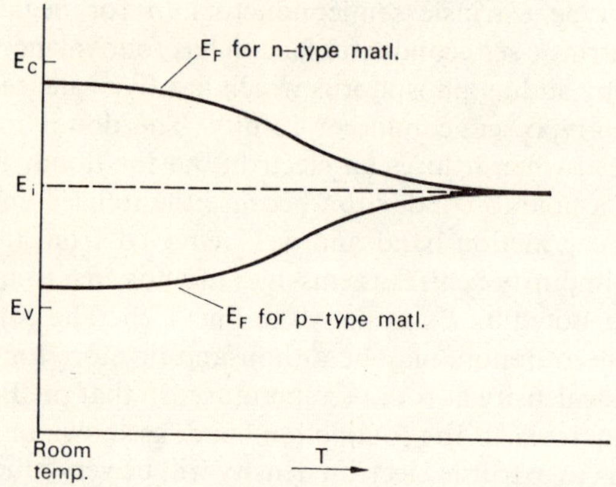

FIG. 1.3. Change of Fermi energy level with temperature.

1.5. The *p–n* Junction

A semiconductor may be doped extrinsic *n* in one region and extrinsic *p* in an immediately adjacent region, and the narrow region in which the changeover occurs forms a *p–n* junction. Such junctions may be formed by growing them into the semiconductor crystal, or by an alloying process, or by diffusion. The latter is by far the most common in integrated circuits and is discussed in detail in Chapter 2.

When a junction is formed, carriers diffuse across it, electrons from *n* to *p* and holes from *p* to *n* until thermal equilibrium is established by means of a potential barrier resulting from the un-neutralized donors and acceptors in the junction region. As mentioned in § 1.3, the Fermi level aligns itself throughout the material under these conditions, and the E_c and E_v levels maintain their positions with respect to E_F in both *p* and *n* regions. The result is that the E_c and E_v levels are raised in the *p*-region relative to the *n*-region as shown in Fig. 1.4a. The magnitude of the potential barrier is V_j (or junction potential) which if measured in electron-volts also gives the numerical value of the junction potential (in volts). The junction potential extends over the *depletion* region, Fig. 1.4a, so called because this region is deplete of carriers. Carriers now have to gain at least this amount of energy to cross from one side to the other.

Although a potential difference exists across the junction, this is not a source of emf. The reason is that the external *p–n* surfaces will also reach thermal equilibrium with each other, and in any circuit the potential barriers will oppose, as for example in a closed *p–n–p* circuit. Of course the equilibrium condition can be disturbed at one of the junctions to create a source of emf, as occurs in a solar cell.

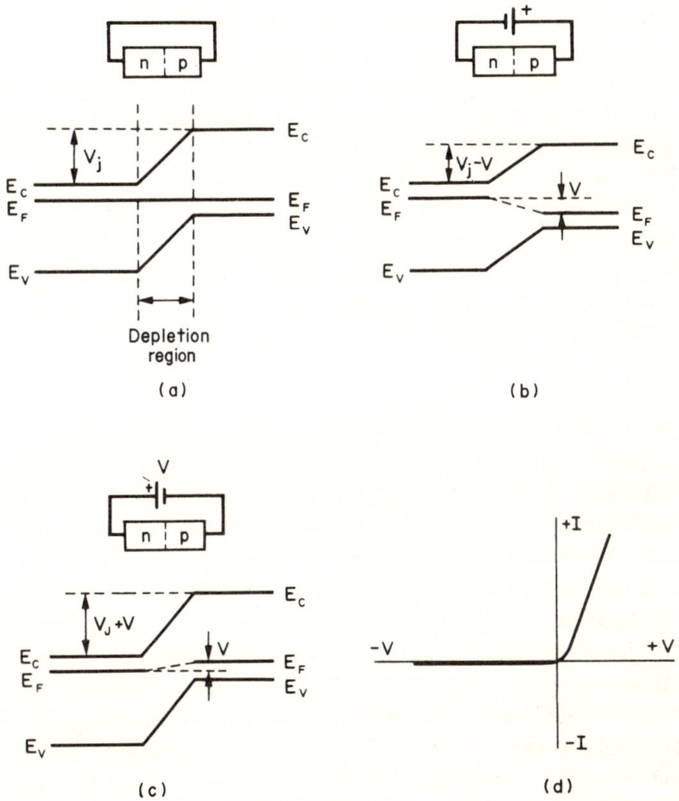

FIG. 1.4. The *p–n* junction energy bands: (a) in thermal equilibrium; (b) forward biased; (c) reverse biased; (d) the V/I characteristic.

Forward bias, Fig. 1.4b disturbs thermal equilibrium and a new *quasi-*equilibrium is established in which the Fermi level on the *p*-side is lowered relative to the *n*-side by an amount in electron-volts numerically equal to the bias voltage. The potential barrier is now reduced to almost $(V_j - V)$ (ignoring any voltage drop in the extrinsic material) and current flows across the junction and around the external circuit. Reverse bias, Fig. 1.4c, increases the barrier potential to $(V_j + V)$ approximately, which prevents current flow. Ideally no current should flow, but there will always in practice be hole–electron pairs created by thermal excitation. Electrons on the *p*-side will form a drift current under the influence of the reverse bias, moving from *p*-side to *n*-side across the junction, while holes on the *n*-side will drift in the opposite direction. The resultant current is known as the *minority* carrier current, and constitutes the reverse current observed in practice for a reverse-biased junction.

It should be noted that when a large forward current flows, the voltage drops in the bulk *p*- and *n*-materials (which up until now have been ignored) always limit the actual forward bias at the junction to a value less than the junction potential, irrespective of the external battery voltage. The voltage–current characteristics for a *p–n* junction for both forward and reverse bias is shown in Fig. 1.4.

1.5.1. THE BIPOLAR JUNCTION TRANSISTOR

An *n–p–n* transistor in rather stylized form is shown in Fig. 3.2a, where the three regions, the emitter, the base, and the collector are connected together to ensure that no external voltages can be applied. The energy-band diagram for this zero bias condition is shown in Fig. 1.5b, and the potential difference diagram in Fig. 1.5c. (It will be recalled that the potential difference results from the charge density of the fixed donor and acceptor centres.) On comparing the energy-band diagram with that of Fig. 1.4 it will be seen that the transistor is in effect two *n–p* junctions connected back-to-back. Thermal equilibrium exists throughout the material as is indicated by the Fermi level being the same throughout, and therefore the net current around the external circuits is zero.

The normal bias conditions for small-signal operation are shown in

Fig. 1.5d. Here it is assumed that the voltage drops across the bulk material are negligible compared with the junction voltages. Thus, the total emitter-base bias voltage V_{EE} appears across the emitter–base junction as forward bias, and the total collector–base voltage V_{CC} appears across the collector–base junction as reverse bias. The effect of these biases is to shift the Fermi energy level as shown in Fig. 1.5e. As a result, the potential barrier at the emitter–base junction is less

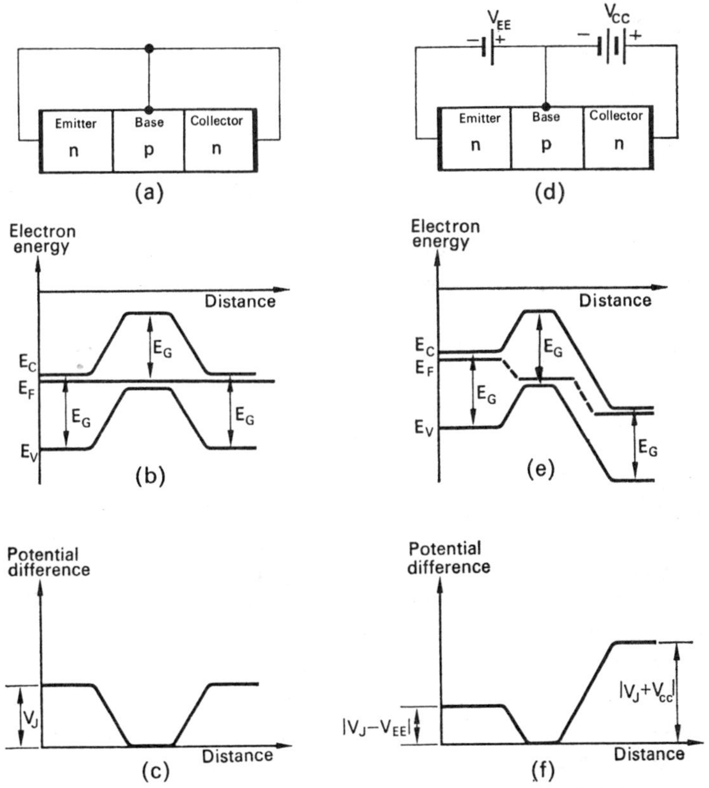

FIG. 1.5. The *n–p–n* transistor: (a) no external bias applied; (b) the energy-band diagram for zero bias; (c) the corresponding potential-difference–distance variation; (d) normal bias conditions; (e) the energy-band diagram for normal bias; (f) the potential-difference–distance variation for normal bias.

than that at thermal equilibrium, and electrons therefore cross the junction from emitter into base. There they are termed *minority carriers* because the base is extrinsically *p*-type.

Holes also move across the junction from base to emitter, but such current is undesirable and it is effectively suppressed by ensuring that the density of electrons on the emitter side is very much greater than the density of holes on the base side of the emitter–base junction. This is achieved in practice by making the impurity doping of the emitter very much greater than that of the base.

The electrons which enter the base from the emitter, being minority carriers on the *p*-side of the reverse-biased collector–base junction, are swept across to the *n*-side (or collector), where they form the external collector current. From the potential difference diagram of Fig. 1.5f it will be seen that the potential difference across the collector–base junction is $(V_J + V_{CC})$.

In addition to there being minority carriers in the base due to emitter injection, minority carriers will be thermally generated as described in § 1.5 (where they accounted for the reverse current of a reverse-biased *p–n* junction). These thermally generated minority carriers also give rise to a reverse current across the collector–base junction, which is termed the *collector saturation current.* It will be seen that the current due to the emitter injection adds directly to the collector saturation current, and in fact the action of the transistor might be thought of as the emitter injection increasing the normal saturation current of the reverse–biased collector–base junction. Modulation of the emitter–base potential by an externally applied signal results in a corresponding signal variation of the collector current, and as the potential difference across the collector–base junction is relatively high compared with that across the emitter–base junction, signal amplification is possible.

In the ideal situation, all the electrons injected by the emitter would reach the collector. In practice, some recombination of electrons with holes takes place as the electrons diffuse across the base. In order to maintain charge neutrality in the base, holes must be injected from the external bias battery V_{EE}, to make up for the holes lost through recombination, and this accounts for the external base current. Normally the base current is a small, approximately constant, fraction of the emitter–collector current (e.g. of the order of 1/50) and thus if the input signal is applied to the emitter–base junction directly via the base current, the signal component of current at the collector output will be amplified (approximately 50 times if the fraction is 1/50).

In some switching circuits the bipolar transistor is operated in what is termed the saturation region (not to be confused with saturation current; see Fig. 4.4), in which the collector–base junction also becomes forward biased. This results in a minority carrier build up on both sides of the collector–base junction which can adversely affect switching speed. This is discussed further in § 4.3.

1.6. The Metal–Semiconductor Junction

The metal–semiconductor (MS) junction is important for two reasons. First, it is required to provide a good electrical contact between wire circuits and semiconductor devices; secondly, certain types of MS junctions exhibit rectifying properties, this effect being utilized in Schottky diodes. The behaviour of a MS junction may be explained also in terms of the energy band diagrams. For the purpose of illustration, an *n*-type semiconductor will be assumed, the energy band diagram being as shown in Fig. 1.6a; this includes an energy term called the work function of the semiconductor E_{ws}. The semiconductor work function is the energy difference between the Fermi level and the zero energy level. In a metal, the energy levels up to the Fermi level E_F may be assumed completely filled, and empty above this. The work function for the metal E_{wm}, is the energy required to remove an electron from the Fermi level to the zero level, as shown in Fig. 1.6a.

When the metal–semiconductor junction is formed, the Fermi level must align in thermal equilibrium, just as in the *p–n* junction. If the work function in the semiconductor is less than that in the metal ($E_{ws}<E_{wm}$), this means that the conduction electrons in the semiconductor are at a higher energy level than the electrons in the metal; electrons will therefore diffuse from the semiconductor to the lower energy level in the metal. A depletion region, with associated potential barrier will form to limit this diffusion process, and the energy bands will bend upwards as shown in Fig. 1.6b. Note that E_F remains level so that the E_c—E_F gap widens, indicating that at the surface the material becomes less *n*-type. A potential barrier V_b is formed as a result of the

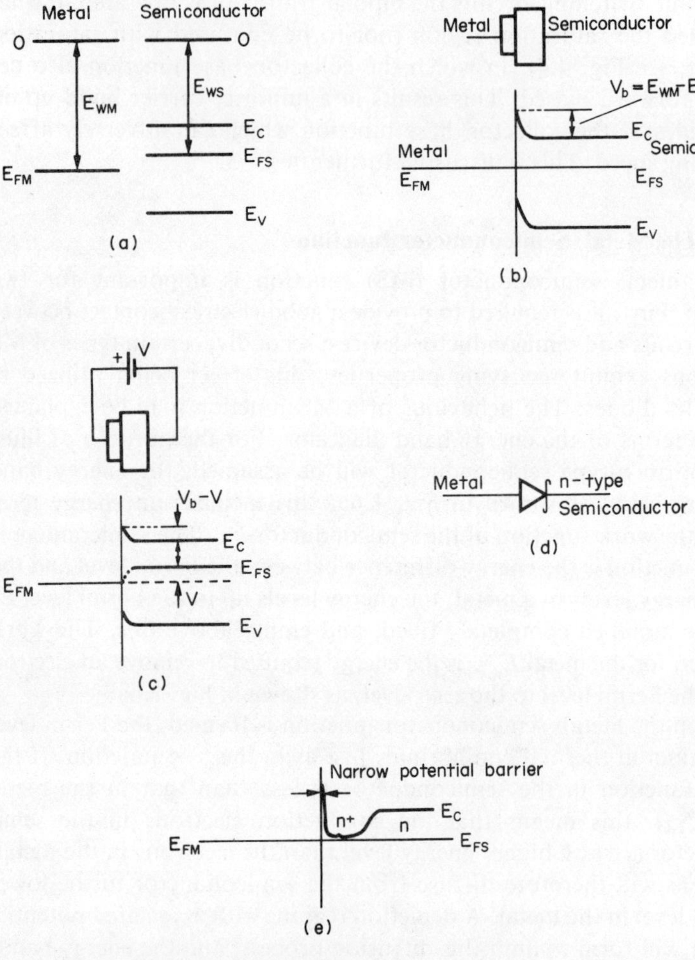

Fig. 1.6. (a) Energy bands in metal and semiconductor before contact. (b) Formation of rectifying contact (E_{wm}>E_{ws}) after thermal equilibrium has been established. (c) Forward bias applied. (d) Symbol for the Schottky Barrier Diode. Diode arrow shows direction of conventional current flow. (e) Formation of ohmic contact through n^+ diffusion.

negative charge on the metal surface and the positive (donor) charge distributed some way into the semiconductor. The height of the potential barrier can be altered by application of external bias (in a similar manner to that for the *p–n* junction shown in Fig. 1.4). Figure 1.6(c) shows the conditions for forward bias. Reverse bias will increase the barrier height to $V_b + V$. Diodes formed in this way are known as Schottky Barrier Diodes (SBDs) and an important feature is that conduction is by *majority carriers.* The conventional symbol is shown in Fig. 1.6(d). Some applications are discussed in sections 4.6, 4.7, and 11.5. Rectifying contacts (or SBDs) can be made with metal on *p*-type semiconductor if the work function for the semiconductor is greater than that for the metal, but the technology here is not as advanced as that for the *n*-type discussed. Also it should be noted that in practice, surface energy states resulting from impurities and defects at the interface may swamp out the rectifying action, so that in the manufacture of these diodes, clean, defect-free surfaces are required.

Where a normal, or non-rectifying, contact is required, usually referred to as an ohmic contact, the work function of the *n*-type semiconductor should be greater than that of the metal, such that the conduction band bends downwards to overlap the Fermi level in thermal equilibrium. Electron transfer across the interface is then easily accomplished for either polarity of external voltage. Ohmic contacts can also be made where the metal forms a true alloy with the semiconductor. Gold welded to *p*-type Ge forms an Ohmic contact while gold welded to *n*-type Ge forms a diode (known as a gold bonded diode). A common method of obtaining an ohmic contact is to heavily dope the semiconductor at the contact region. For an *n*-type semiconductor this is indicated as n^+ and the energy band diagram is sketched in Fig. 1.6(e). Electrons tunnel through the narrow potential barrier (bearing in mind that the quantum-mechanical model of the electron is required in which the electron is spread out in space rather than localized as a point particle). Metal-p^+-p type semiconductor contacts can be made in a similar manner.

1.7. The Metal–Insulator–Semiconductor (MIS) Junction

Most widely used is the Metal–Oxide–Silicon (MOS) structure found in the MOS transistor (MOST), and in charge coupled devices (CCD) described in Chapter 6. Again, for purposes of illustration, an *n*-type semiconductor is assumed, and the energy band diagram is as shown in Fig. 1.7a for the condition where no external bias is applied, and no surface energy states exist at the semiconductor–insulator interface. As before, the Fermi level must align throughout the structure under conditions of thermal equilibrium. Since no transfer of charge can take place across the insulator (note that the insulator conduction band is empty, and the valence band is well below the Fermi level) the energy levels in the semiconductor will remain level and shift together relative to the Fermi level in the metal; this will result in a displacement in the zero levels relative to each other.

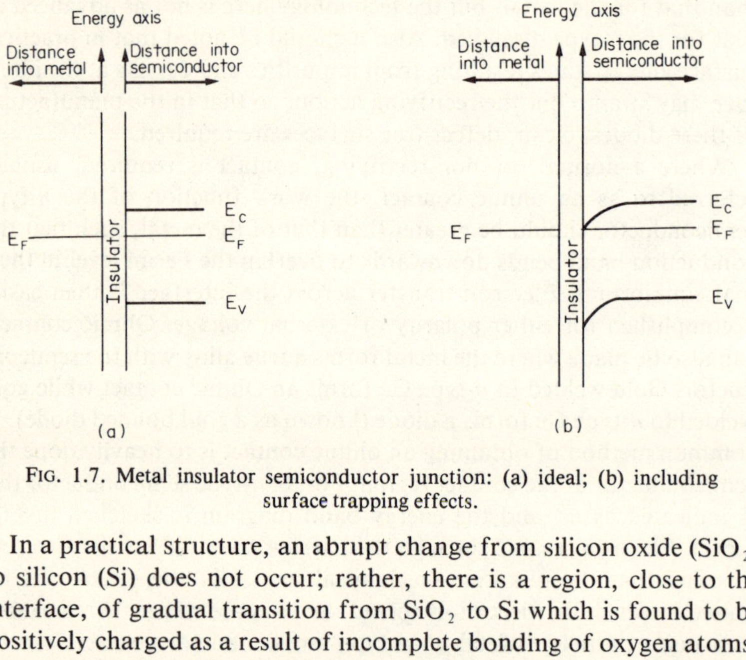

FIG. 1.7. Metal insulator semiconductor junction: (a) ideal; (b) including surface trapping effects.

In a practical structure, an abrupt change from silicon oxide (SiO_2) to silicon (Si) does not occur; rather, there is a region, close to the interface, of gradual transition from SiO_2 to Si which is found to be positively charged as a result of incomplete bonding of oxygen atoms. The positive charge is assumed to reside at the interface on the SiO_2 side, and it induces an equal and opposite charge in the semiconductor at the interface. This effect critically affects device performance, as discussed in Chapter 6. The energy band diagram taking into account this charge dipole is sketched in Fig. 1.7b.

1.7.1. THE METAL–INSULATOR–SEMICONDUCTOR TRANSISTOR

By applying an external potential to an MIS structure the carrier density at the surface of the semiconductor can be altered, this in turn altering the surface conductivity. Control of surface conductivity in this way forms the basis of the field effect transistor. The metal electrode is termed the *gate*, and Fig. 1.8 illustrates how the energy bands are altered when a voltage is applied between gate and semiconducting channel. Again, the rule to follow is that the Fermi level remains constant for equilibrium. An *accumulation layer* is created at the surface of the semiconductor when the electron density is increased there (as for example occurs when the *n*-channel depletion mode transistor of Fig. 6.3 is operated in the enhancement mode). In order to accommodate this increased density, the energy levels are bent downwards with respect to the Fermi level, as shown in Fig. 1.8a. In effect, the external gate potential lowers the potential energy at the surface of the semiconductor, and as the natural tendency is for the lowest available energy levels to be filled, electrons will move to these lower energy states at the surface. The opposite effect occurs when electrons are repelled from the interface as, for example, occurs when the gate potential of the *n*-channel transistor is made negative. The potential energy of available states at the surface is now increased and these are more difficult to fill, with the result that the electron density is reduced. At the same time, minority carriers, holes, in the bulk semiconductor will be induced to the surface. The surface layer under these conditions is termed a *depletion* layer. By increasing the applied voltage the hole density at the surface can be made greater than the electron density, and the surface is then said to be *inverted*. This corresponds to the conditions in a *p*-channel enhancement mode transistor.

Field effect transistors are discussed in more detail in Chapter 6.

1.8. Effective Mass of a Carrier

When an electron in free space is subjected to an electrical field, it moves according to the normal laws of motion, the mass remaining constant. In a solid, it is found that for the laws of motion to apply an

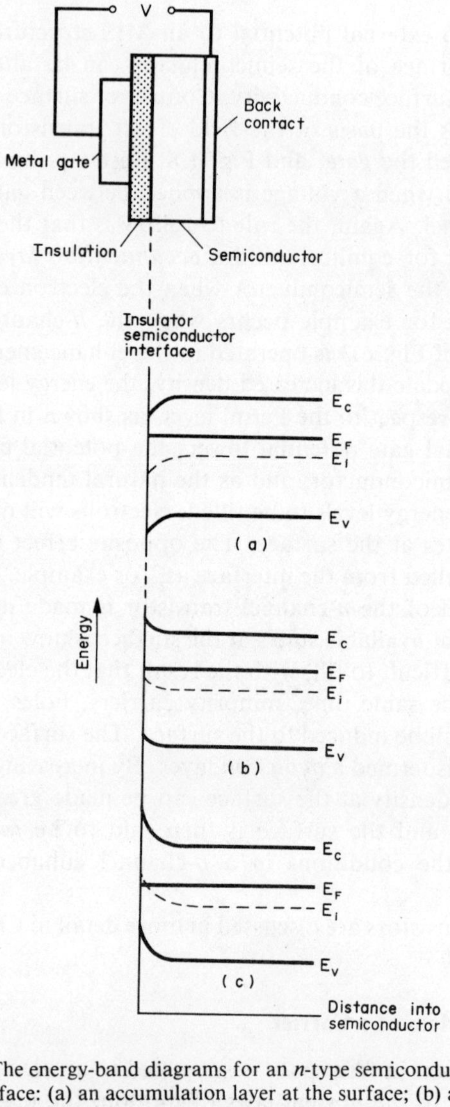

FIG. 1.8. The energy-band diagrams for an *n*-type semiconductor as modified by an interface: (a) an accumulation layer at the surface; (b) a depletion layer at the surface; (c) an inversion layer at the surface.

effective mass must be assumed for the electron which in general is different from the free space mass. The effective mass varies in a complicated way as a function of the electron energy. This peculiar behaviour results from the regular variation of potential energy with position throughout the atomic lattice. Practically, this means that an electron in a solid may have an effective mass which ranges from being heavier than, to being lighter than, the free electron mass. One consequence of this is that electrons may be transferred from a light-mass energy band to a heavy-mass energy band which can result in current fluctuations at microwave frequencies; this is the basis of the Gunn oscillator (see Chapter 11).

A hole may also have an effective mass which differs from the equivalent free-space mass (equal to the electron free-space mass), and one has then the interesting situation of having heavy holes and light holes!

The usual symbol for the effective mass is m^* and subscripts may be added for electrons and holes respectively thus: m_e^* and m_h^*.

1.9. Carrier Mobilities

A carrier (electron or hole), when drifting under the influence of an applied electric field, will collide with imperfections in the atomic lattice and as a result it will attain a limiting velocity (compared to an electron in free space which undergoes continuous acceleration when subjected to an electric field). In the simplest case the average drift velocity is directly proportional to the applied field strength, the constant of proportionality being known as the *drift mobility*, (symbol μ). The equation relating drift velocity $\mathbf{v}$ to electric field $\mathbf{E}$ is:

$$\left.\begin{aligned} &\text{electrons:} \quad \mathbf{v} = -\mu_e \,.\, \mathbf{E} \\ &\text{holes:} \quad \mathbf{v} = \mu_h \,.\, \mathbf{E}\,. \end{aligned}\right\} \qquad (1.1)$$

The subscripts e and h are for electrons and holes respectively, and the negative sign for electrons is to show that the velocity is directed opposite to the electric field direction because the electrons are negatively charged. Both μ_e and μ_h are positive numbers, and are inversely proportional to the effective mass of electron and hole, respectively.

TABLE 1.1.
PROPERTIES OF SOME SINGLE CRYSTAL SEMICONDUCTORS AT ROOM TEMPERATURE (290°K)

Material	Symbol	Drift mobility (cm^2/volt—sec.)		Effective mass ratio		Energy band-gap	Intrinsic carrier-density
		μ_e	μ_h	m_e^*/m	m_h^*/m	$E_{(G)}$eV	(n_i) cm^{-3}
Germanium	Ge	3800	1800	0.17	0.042 0.34	0.7	2.2×10^{13}
Silicon	Si	1450	500	0.31	0.16 0.52	1.15	1.4×10^{10}
Tellurium	Te	1170	560	0.68	0.91	0.34	9.3×10^{15}
Gallium-arsenide	Ga As	9000	400	0.07	[a]	1.35	1.4×10^{6}
Indium-antimonide	In Sb	70000	1000	0.013	0.18	0.18	1.35×10^{16}

[a] Data not available.

Some values are listed in Table 1.1. Mobility is seen to have dimensions of Velocity/Electric Field, the units most commonly used being cm^2/Volt-sec.

High mobility is desirable for good high-frequency performance and fast switching performance of devices. The values listed in Table 1.1 apply within the bulk of a single crystal material. When the carriers are moving near the surface, the mobility tends to be reduced because of additional collisions at imperfections near the surface, and carrier trapping effects. (A trapped carrier is one which is held at an energy level within the forbidden band and so cannot contribute to conduction. These levels within the otherwise forbidden band occur mainly because of the discontinuous nature of the insulator–semiconductor interface, and they are confined mostly to the surface of the semiconductor, i.e. they do not extend far into the bulk of the semiconductor.) Thus, when dealing with the motion of carriers in the surface channel of a MOST, the idea of *surface mobility* must be used, and this, in general, will be lower than the bulk mobility. In practice, it is found easier to measure the change in channel conductance as a function of applied gate charge, and this gives rise to the definition of *field effect mobility* which is closely related to, but not necessarily equal to, the surface mobility. Only when surface trapping effects are absent are the two equal.

The field effect mobility μ_{FE} is defined as the change in surface conductivity as a function of gate charge per unit area, and therefore does not relate to the movement of any specific group of carriers. Since drain current is proportional to channel conductivity, and gate voltage to gate charge, the mutual conductance g_m of the device is by *definition,* directly related to μ_{FE}, as shown in eqn. (6.1.).

1.10. Crystalline Structure

For most technological applications of semiconductors, *single crystal* semiconductor material is required. In single crystal material the atoms are arranged in a perfectly regular pattern which extends completely throughout the material without breaks or dislocations. The space pattern for the atoms is represented by a *crystal lattice* which is defined as a three-dimensional arrangement of points in which every point has exactly the same environment. (The atoms need not be situated *at* the points of the lattice but can be symmetrically placed with respect to these.) At the surface of a solid the environment of a lattice point obviously changes, i.e. the surface represents a gross dislocation

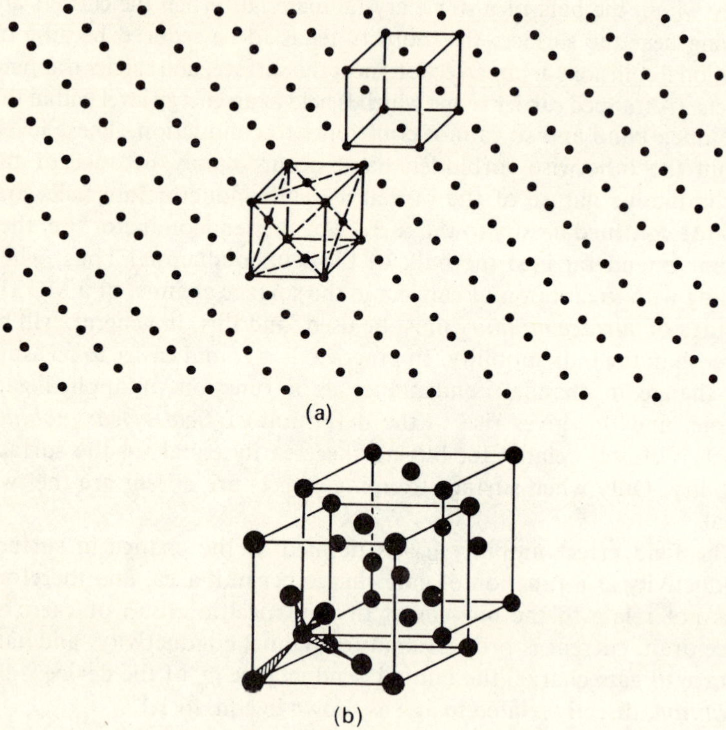

FIG. 1.9. (a) A face-centred cubic (f.c.c.) lattice; (b) the silicon lattice, which can be represented by two intermeshing f.c.c. lattices. Nearest neighbour bonding is shown for one group of atoms.

of the lattice symmetry, and it is for this reason that surfaces play such an important part in the behaviour of solid-state devices.

Figure 1.9a shows a *face-centred cubic* lattice (f.c.c.) in which a unit cell is outlined. The name arises because a lattice point occurs at the centre of each face of the unit cube, as shown.

Single-crystal silicon consists of two f.c.c. lattices intermeshing in such a way that one is displaced by one-quarter way along the body diagonal of the other, as shown in Fig. 1.9b. Because of this arrangement each and every atom has four nearest neighbours. A very strong bond is formed between nearest neighbour atoms through sharing of valence band electrons, and this type of bonding is known as *covalent bonding*. The bonding is strongest when each bond (between any two nearest neighbour atoms) has exactly two electrons. This condition exists in silicon, in which each atom has four valence band electrons. Any one atom will share four valence electrons, one from each of its nearest neighbour atoms, while it itself contributes its four valence electrons. Therefore a total of eight electrons make up the four nearest neighbour covalent bonds, i.e. an average of two electrons per bond.

Covalent bonding is also found in other solids such as diamond (an insulator) and germanium (a semiconductor).

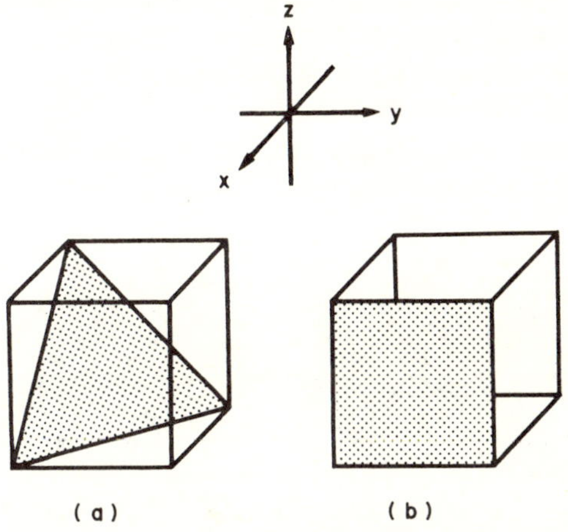

FIG. 1.10. (a) The (111) surface, and (b) the (100) surface of a cubic crystal.

Of great importance in the manufacture of integrated circuits is the silicon surface selected for processing (described in the following chapter). The two most widely used surfaces are the (111) and the (100), these being illustrated in Fig. 1.10a and b, in relation to the (xyz) axes of the f.c.c. crystal. The numbers used to specify a plane or surface in a crystal are termed *Miller Indices,* and for any *cubic crystal,* the (*xyz*) coordinates of the surface normal are numerically equal to the Miller Indices for that surface.

Different surfaces exhibit different properties. Where alloyed junctions are to be formed (see § 2.7) alloying into the (111) surface produces a flat junction. Where MOS type structures are formed by growing silicon dioxide (SiO_2) on the silicon surface, it is found that the trapped oxide charge discussed in connection with Fig. 1.7b is considerably less for the (100) surface, compared with the (111) surface; use is made of this to reduce the threshold voltage in MOS transistors, and is discussed further in § 6.4.

Epitaxial crystal growth occurs when a crystal lattice is continued without a break into a layer grown on a parent crystal. (Epitaxial means around the same axis, in this case crystal axis.) A sketch of an epitaxial layer is shown in Fig. 1.11. Epitaxial growth is of great importance in semiconductor technology as it enables material characteristics to be precisely controlled, and, for example, it is possible to grow an *n*-type layer onto a *p*-type substrate in this way.

In other applications *polycrystalline* materials may be used, as, for example, in thick- and thin-film circuits. A polycrystalline material is composed of a number of single-crystal regions, called grains, packed

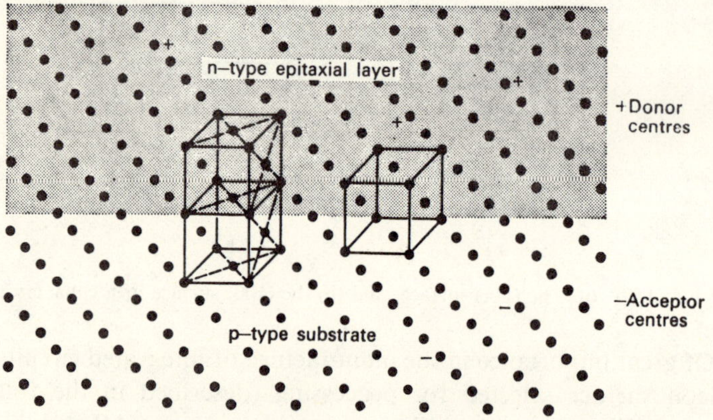

FIG. 1.11. An epitaxial *n*-type layer on a *p*-type substrate.

together in an irregular fashion. Grain boundaries are defects in a material which can have the effect of reducing carrier mobility. However, free carrier density in a polycrystalline semiconductor is very considerably increased and the conductivity can approach that of a metal. Use is made of the near metallic nature of "polysilicon" and the fact that silicon processing steps can be used with it, to fabricate gate electrodes in MOS transistors (see Chapter 6).

Where there is a complete absence of crystalline structure (or at most this extends only over a small group of atoms at a time), the material is said to be *amorphous*. Most glasses, for example, are amorphous. Recently, semiconducting glasses (which are in effect amorphous semiconductors) have been developed which appear to exhibit useful switching properties.

Some of the properties for intrinsic semiconductors at room temperature are summarized in Table 1.1. These values are taken from various published results (see, for example, R. A. Smith, *Semiconductors*, Cambridge Univ. Press, 1964) and are included to give an idea of typical values. As Table 1.1 is only intended as a guide to typical values, where necessary an average of various published values has been used. The effective mass ratio is the ratio of the effective mass (see § 1.8) of a carrier to the free-space mass of an electron, which is 9.1×10^{-31} kg. It will be noted that two effective hole masses are observed for silicon and germanium.

CHAPTER 2

Processing of Silicon Devices and Circuits

2.1. Introduction

The complete processing of a silicon slice to produce devices or circuits involves a number of quite distinct stages such as photo-resist application, chemical etching, and diffusion of impurities into the silicon. The utmost care and cleanliness must be observed at all stages and as a result most of the work is carried out in clean-rooms. These are specially designed rooms in which the air is filtered to remove dust particles and the room pressure is maintained slightly positive with respect to the outside ambient pressure to prevent accidental entry of unfiltered air. Operators in the room must wear special clean-room clothing which includes shoes, hats and gloves. Gas and water supplies for processing purposes must be pure, and deionized water is invariably used.

A special problem arises in the manufacture of insulated-gate transistors. Electrostatic charge generated by nylon clean-room clothing can cause gate-breakdown of these transistors, and to prevent charge build-up the operators may have to be chained to their work trolleys by means of metal bracelets!

2.2. Preparation of Artwork

The artwork for a device or circuit starts with the large-scale layout of the particular patterns required. The scale factor can range from 200 : 1 to 1000 : 1, a typical value being 250 : 1. Suppose, for example, that two electrodes have to be spaced by 10 μm in the actual circuit.

These would be shown spaced by $10 \times 250\ \mu m$ or 2.5 mm, on the large-scale artwork for a scaling factor of 250 : 1. Close tolerances can be achieved on the large-scale layout and maintained on the reduced image.

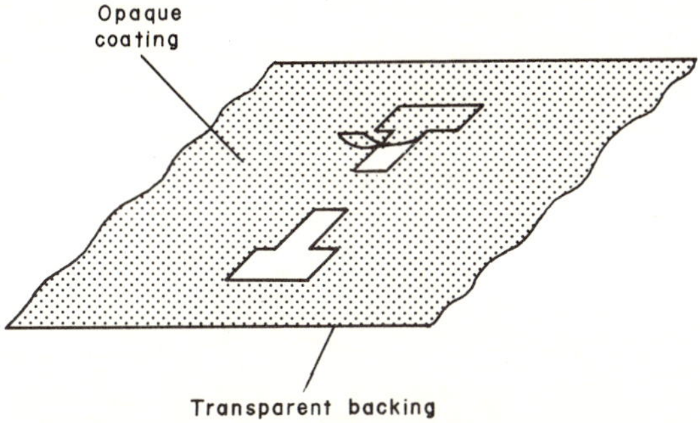

FIG. 2.1. (a) Scribe and peel material for large-scale artwork.

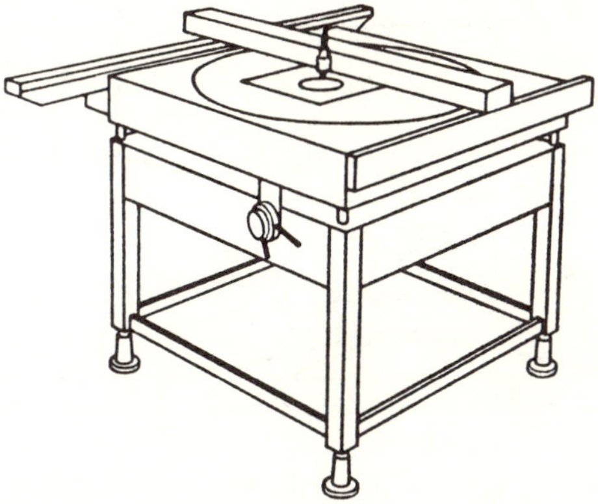

FIG. 2.1. (b) A co-ordinatograph. (Courtesy: Consul & Mutoh Ltd.)

There are a variety of ways in which the large-scale artwork can be prepared. One method utilizes peel-coat material, sketched in Fig. 2.1a. This consists of an opaque coating on a transparent backing, and in use the coating is cut and peeled leaving the desired pattern showing through the transparent backing. The cutting has to be precisely controlled and this is carried out on a special machine known as a co-ordinatograph (Fig. 2.1b).

The large-scale artwork is photographically reduced, usually in two stages, the first stage giving a 25 : 1 reduction and the second a 10 : 1 reduction (for a total reduction of 250 : 1). For the first reduction the cut and peeled artwork is laid on an evenly illuminated screen and photographed. Both the camera and the screen carrying the artwork are mounted on a shock- and vibration-free base. In some cases this may weigh in the order of a ton and is spring-mounted. Along with these requirements the camera lens must be capable of the high resolution required and the film plate must have the necessary dimensional stability.

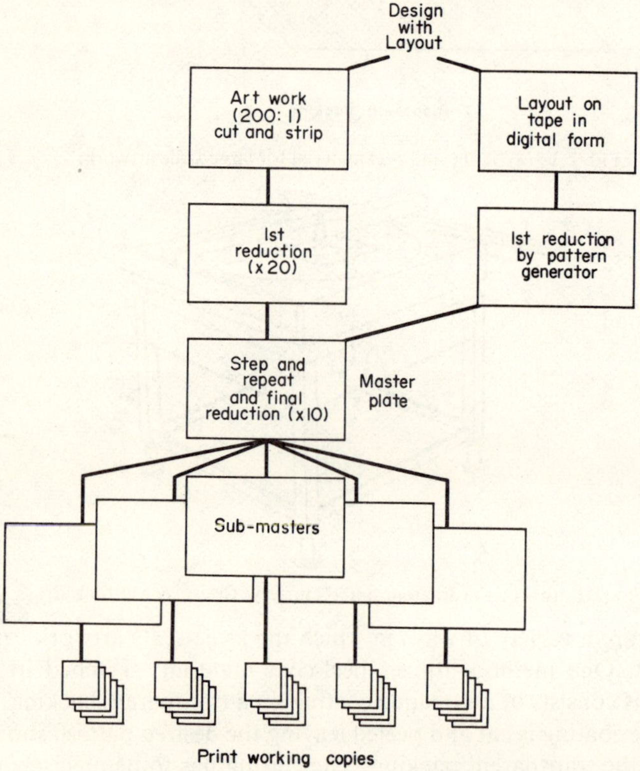

FIG. 2.2. Mask making sequence for production copies (Courtesy: *The Radio & Electronic Engineer*, and M. Smollett).

In the second reduction a step-and-repeat procedure is followed. Here the image is reduced to its final size and at the same time projected onto the final photographic negative. The negative is shifted, or stepped, by a precisely controlled amount and exposure repeated. (The actual exposure may be achieved by having a flashing light synchronized with the movement of a co-ordinate table carrying the negative.) The final photographic negative carries an array of images for the device or circuit to be made. Copies are made of the negative and used in the processing steps for the silicon, as the master negative is expensive to make. It is stored carefully for copying only.

For complex circuits such as those utilized in Large Scale Integration (LSI), computer aided design is absolutely essential, and the cut-and-peel approach for circuit design is not practicable. Instead, the circuit layout information is prepared in the form of a digitally-encoded tape which in turn is used to control a pattern generator. The pattern generated is then reduced optically onto a photographic plate, and copies made as in the previous method. Both methods are outlined in Fig. 2.2.

2.3. Photo-sensitive Resists

Photo-sensitive resist is a liquid solution of resin in an organic solvent, which when exposed to ultraviolet light changes its molecular structure. With *negative* photo-resist, regions that are exposed to ultraviolet light harden (the molecules polymerize) and form a strong chemical resistant cover when developed. The unexposed regions dissolve away in the developing stage. With *positive* photo-resist, the reverse happens, the exposed regions being removed in the developing stage and the unexposed regions forming the tough chemical resistant cover. Figure 2.3a shows a negative photo-resist layer being exposed to ultraviolet light through a photographic mask, and Fig. 2.3b shows the developed photo-resist pattern that results from this. Figure 2.3c shows a positive photo-resist layer being exposed to ultraviolet light through the same photographic mask, and Fig. 2.3d the resulting pattern that is developed in this case.

The application of the photo-resist is one of many critical steps in the process, as at this stage dust particles or lumps in the photo-resist

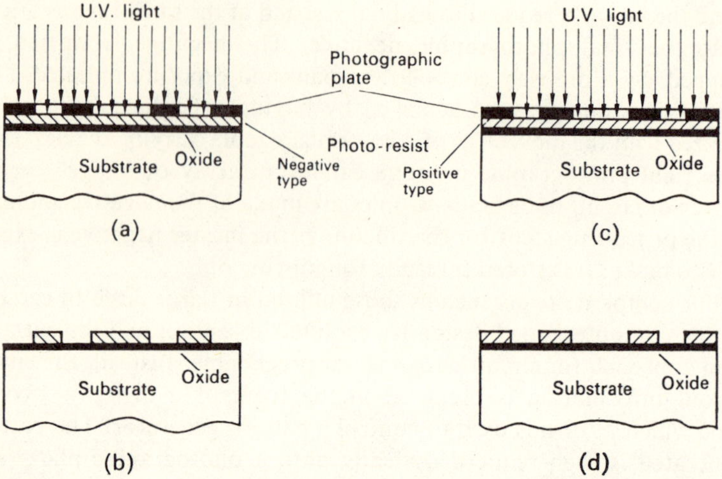

FIG. 2.3. the photo-sensitive resist stage: (a) exposure of *n*-type photo-resist; (b) the developed image; (c) exposure of *p*-type photo-resist; (d) the developed image.

can cause device failure (either immediate or in the long-term). The resist is filtered immediately before being applied to the substrate to be coated. When applied, the substrate is spun causing the resist to flow evenly over the surface. The thinner the coating of photo-resist, the better the image definition that can be obtained, but obviously the chemical resistance will decrease with decreasing thickness. With spinning speeds of the order of 12,000 r.p.m. coating thicknesses of the order of 5000 Å can be obtained. Prior to applying the resist it is essential that the substrate be clean and dry.

Following the application of the photo-resist, the coating is allowed to dry, normally at room temperature for about 5 min and then by applying heat at about 60°C for about 5 min. This is followed by a prebake cycle typically fifteen minutes at 100°C. Since the coating is photo-sensitive the work must be carried out under safe lighting, e.g. amber light, and under clean-room conditions.

During the actual exposure to the ultraviolet light the photographic negative is held in very close contact to the photo-resist layer, usually in a vacuum frame. After exposure the resist is put through a post-bake cycle similar to the pre-bake cycle.

2.4. Masking and Diffusion

The photo-resist mask cannot be used directly in the diffusion process because this is carried out at a high temperature which would completely destroy the resist coating. Instead, the silicon wafer is oxidized before the photo-resist is applied. The oxide is then chemically etched through the photo-resist mask the etchant attacking only the oxide, and not the hardened resist coating.

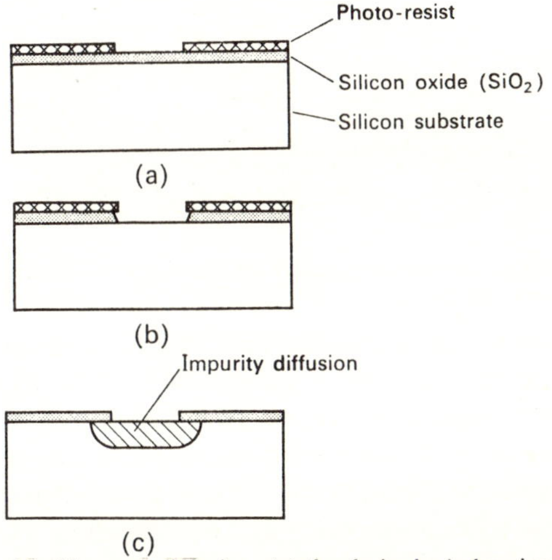

FIG. 2.4. Masking and diffusion: (a) the desired window image developed in the photo-resist; (b) the window etched in the silicon dioxide layer; (c) impurity diffusion through the window.

The oxide is formed by placing the cleaned silicon wafer in a steam atmosphere at about 1200°C for 1 hr. The resulting oxide is silicon dioxide (SiO_2) and is between 1 and 2 μm thick. Here again great care must be taken to prevent impurities from entering the oxide.

Assuming that a simple window has to be etched into the oxide, the oxidized silicon slice with the photo-resist mask would appear as shown in Fig. 2.4a. The wafer is then subjected to an etching bath (a typical etchant for SiO_2 being a solution of ammonium fluoride and hydrofluoric acid) which attacks the oxide but not the silicon nor the resist coating. A window is therefore etched in the oxide as shown in Fig. 2.4b. The wafer is again cleaned, including removal of the photo-resist, and placed in a furnace at about 1100°C which contains in vapour form the desired impurity. This diffuses through the window in the oxide into the silicon to form the required *p*- or *n*-type region, as shown in Fig. 2.4c.

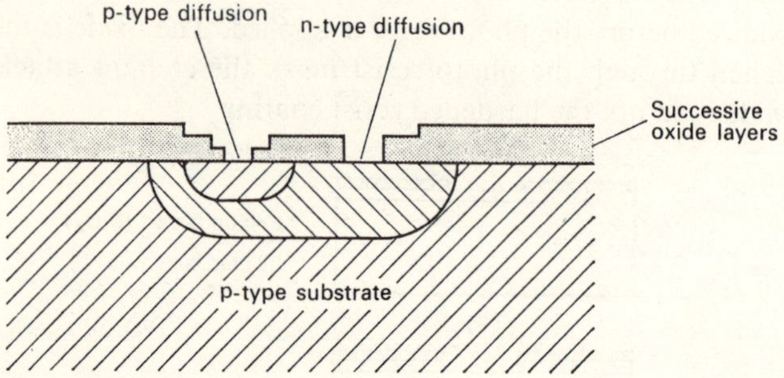

FIG. 2.5. Successive *p*- and *n*-type diffusions in a *p*-type substrate.

In practice a number of diffusions are usually required and these have to be critically aligned with respect to each other. This calls for the use of a precision optical viewing system, coupled with precision mechanical movement of the work-piece to enable successive masks to be aligned with respect to the diffusion regions already formed. Alignment of masks has to be, typically, within a fraction of a micron. Figure 2.5 shows a section of a *p*-type silicon wafer which has been subjected to first an *n*-type diffusion (arsenic or phosphorus being typical dopants) followed by a *p*-type diffusion (boron and gallium being typical dopants). Further oxidization and etching is required to form windows for the connections and the oxide layer formed at this stage is left on as a protective coating. The structure shown in Fig. 2.5 is the basic structure required for a single *p–n–p* transistor.

Although silicon dioxide is widely used as a masking agent, its one serious disadvantage is that it allows certain ions to move easily through it, the sodium ion being one of the most troublesome. With insulated-gate transistors in which the masking oxide is also used as the gate insulator, any ions present in the oxide usually result in instability of the transistor's characteristics. To prevent the diffusion of ions in the gate-insulator, a double layer insulator is frequently used in which one of the layers consists of silicon nitride (Si_3N_4). Alternatively, aluminium oxide (Al_2O_3) may be used as an ion barrier instead of the nitride, but here the aluminium oxide must be deposited on top of the silicon oxide by means of vacuum evaporation (see § 8.2).

2.4.1. GOLD DIFFUSION

Gold diffused into silicon creates localized energy levels within the otherwise forbidden energy gap which enables holes and electrons to recombine faster than if they were to rely on direct recombination across the band-gap. This is applied in some cases to speed up switching (see § 4.3). Gold diffusion is achieved by first evaporating a gold layer on the back of the silicon slice and then placing this in a diffusion furnace for a few minutes. Gold diffuses easily through silicon, and some manufacturers rely on this occurring during emitter diffusion stages for transistors, while others have a separate step for the gold diffusion at a later stage in the manufacturing process.

2.5. Metallization

Where an interconnecting conductor pattern (referred to as the metallization) has to be deposited on the top surface of the silicon integrated circuit, or where other thin-film circuits are to be made, positive or negative photo-resist can be used; in addition, the thin-film material can be removed either by etching or, alternatively, by a lifting process. These various possible methods are illustrated in Fig. 2.6, where, for clarity, a very simple interconnecting pattern is assumed. As shown, the metallization is required to connect together regions *A* and *B* on the silicon circuit, while avoiding regions *X* and *Y*.

In the etching process (i.e. branching to the left in Fig. 2.6) the silicon is coated on the top surface first with a metallic layer, usually aluminium. This is deposited by vacuum evaporation (see § 8.2) to a thickness of about 1 μm. Next a coating of photo-resist is applied and processed, as described in § 2.3. The appropriate mask is selected as shown in Fig. 2.6, depending on the type of photo-resist used. Note that one mask is the negative of the other, and both masks may be either normal photographic emulsion, or may be chrome, on glass. These masks are known as the working plates. The photo-resist is exposed and developed as described in § 2.3, and the aluminium etched to form the desired pattern. The hardened photo-resist protects the aluminium immediately underneath, apart from slight undercutting.

In the lifting process (branching to the right in Fig. 2.6), the first step is to coat the top of the silicon surface with photo-resist and to develop the pattern in this as shown, again using the mask appropriate to the photo-resist. The aluminium layer is deposited on top of the photo-resist, and subsequent removal of the photo-resist lifts off the aluminium immediately above it leaving the desired pattern on the silicon.

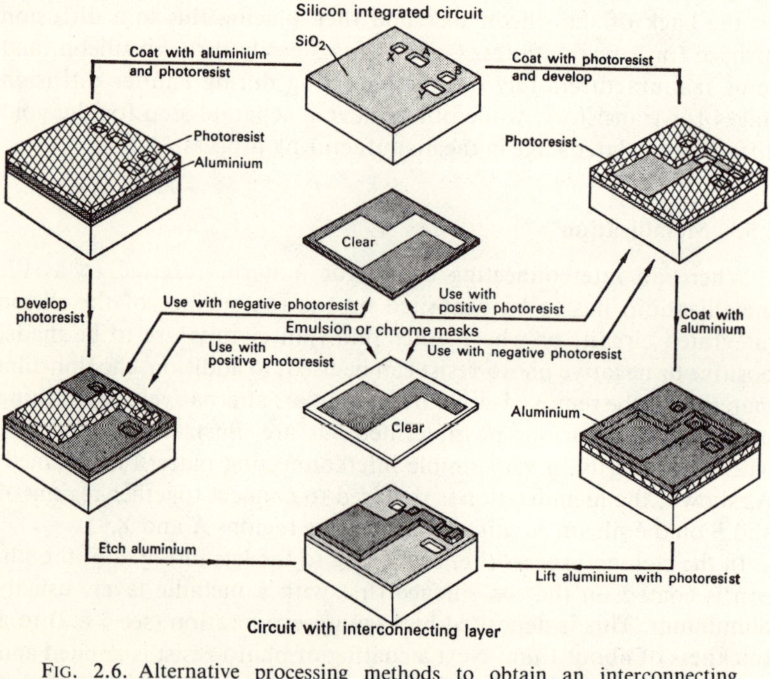

FIG. 2.6. Alternative processing methods to obtain an interconnecting pattern. The left branch shows the etching method, the right branch the lifting method.

The choice of any particular sequence is determined in practice by a number of factors but chiefly by the type of defects associated with either emulsion or chrome masks. A detailed account of the processes and the factors involved in making a choice of method is given by C. T. Plough, W. F. Crevier, and D. O. Davis, Photo-resist process optimization for production of planar devices, published in the *Proceedings of the Second Kodak Seminar on Microminiaturization,* April 1966.

2.6. Epitaxial Growth

The meaning of epitaxial growth in relation to crystal growth is explained in § 1.10. The main advantages of epitaxial growth are: (i) very thin layers (a few microns thick) can be grown on an existing substrate, and these are required for fast switching diodes and transistors;

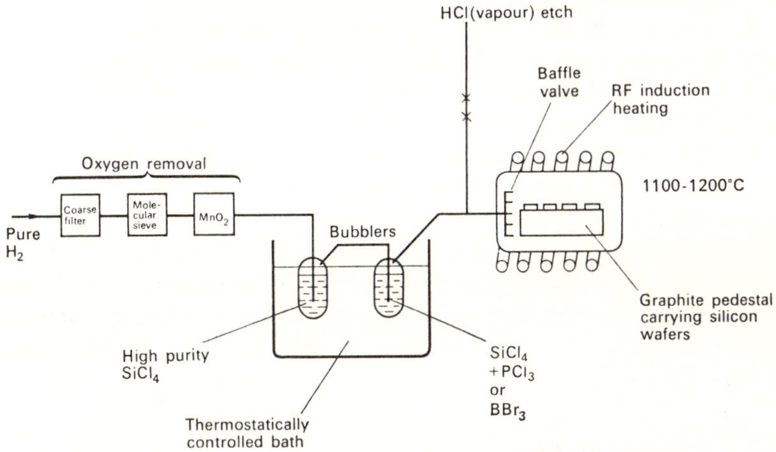

FIG. 2.7. Epitaxial growth on silicon by reduction of silicon-tetrachloride.

(ii) the impurity doping of the epitaxial layer can be precisely controlled throughout the complete thickness of the layer and independently of the main substrate doping. For example, a uniform distribution of impurities throughout the layer can be achieved.

Figure 2.7 shows one method by which epitaxial growth may be achieved on silicon. This is known as a reduction process in chemistry, and the chemical equation describing it is:

$$SiCl_4 + 4H = Si + 4HCl \tag{2.1}$$

i.e. by adding hydrogen (H), to the silicon tetrachloride ($SiCl_4$) and passing the mixture into the reaction chamber, at higher temperatures it is reduced to silicon (Si) plus hydrochloric acid (HCl). In the reaction chamber the silicon in vapour phase settles onto the single crystal wafers on the graphite pedestal, and results in an epitaxial layer growing at a rate of about 1μm/min. The remaining atmosphere in the reaction chamber is hydrochloric acid in the vapour phase and this must be extracted to maintain a constant flow rate of input–output gas. Hydrochloric acid is already present in the system as this is used prior to heating the silicon for the removal of surface contaminants (shown as the HCl etchant inlet on Fig. 2.7), but this source of HCl is closed off during the epitaxial growth stage.

Pure hydrogen must be used, with all traces of oxygen removed, and this is achieved by the filtering system shown in Fig. 2.7. The hydrogen is bubbled through a high-purity silicon-tetrachloride bath to form the mixture given by the left-hand side of eqn. (2.1). The mixture is passed through a further bubbler containing silicon tetrachloride and the desired impurity dopant. Phosphorus trichloride (PCl_3) is the usual dopant for *n*-type impurity, and boron tribromide (BBr_3) is usual for *p*-type doping, the active doping agents being the phosphorus and boron respectively. The atoms of these dopants replace silicon atoms at various sites throughout the crystal lattice, a typical impurity content being 1 in 10^4, i.e. one impurity atom for every 10,000 silicon atoms. The impurity density can be controlled very closely.

2.7. Alloyed Junctions

Although the diffusion method described in § 2.4 is the standard production method presently in use for most devices and integrated circuits, brief mention will be made of the alloy process, as some discrete devices in germanium and silicon are still made by this method. Basically it is very simple. A small pellet of metal is placed on the semiconductor and together they are heated in a non-oxidizing atmosphere until the metal melts and dissolves the region of semiconductor immediately under it, to form an alloy. On cooling the alloyed region recrystallizes as a single crystal continuous with the unmelted semiconductor crystal, but is heavily doped with the metal. p^+-n junctions are formed this way using Aluminium on n-type silicon, and Indium on n-type germanium. As mentioned in § 1.10, alloying into the (111) surface produces an almost flat junction, as sketched in Fig. 2.8. This is because in the f.c.c. crystal (applicable to both Si and Ge; see Figs. 1.9 and 1.10), the atoms are densely packed in the (111) planes while comparatively open channels exist between the planes so that the metal atoms diffuse laterally until they can replace a complete plane. The depth of the junction depends on both the temperature and the time allowed for alloying.

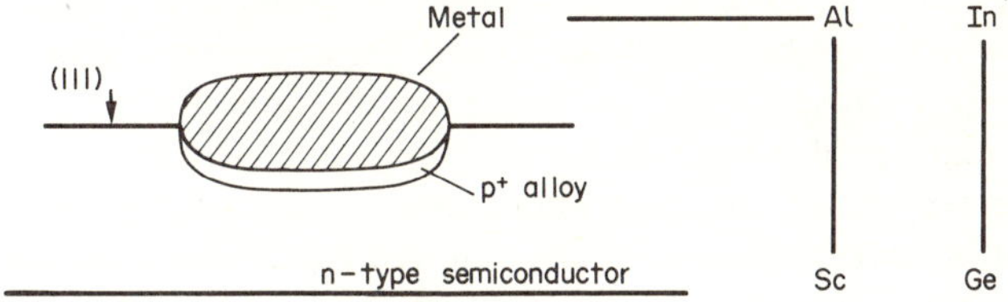

FIG. 2.8. An alloyed junction.

2.8. Summary

It has only been possible here to outline some of the processes employed in the preparation of semiconductor material for use in devices and circuits. It will be apparent that a high degree of perfection is required in the fields of chemistry, photography, optics and mechanical engineering. The electronic circuit design forms only a part of the overall manufacturing process of integrated circuits and the circuit designer should be aware of new possibilities of circuit design opened up by processing techniques, as well as being aware of limitations on circuit design imposed by the processing techniques.

New methods of pattern generation are being constantly explored, some examples being electron-beam machining; irradiation of oxides to produce different etch rates; ion-beam implantation of dopants. It seems likely, however, that the photo-resist method will be the standard production method for many years to come.

CHAPTER 3

Silicon Planar Devices and Integrated Circuits

3.1. Introduction

A large number of devices may be made simultaneously in a silicon wafer using the processing steps described in Chapter 2. A single type of device, which may be a diode or transistor, may be made in large quantities, and subsequently packaged as discrete devices. Alternatively, a number of devices and components may be made together as an integrated circuit, the interconnections being made by means of a metallization layer deposited in the required pattern over the top of the silicon chip. These circuits may contain diodes, transistors, resistors, and capacitors, and are known as silicon integrated circuits (SIC); the name may be further qualified by use of the word *monolithic*, meaning formed in a single block (of silicon).

Two types of transistors are in widespread use, both as discrete components and as components in integrated circuits. The *bipolar* transistor is the conventional transistor and the name arises because both types of carriers, holes and electrons, make up the current flow (two polarities of carrier are involved). The *unipolar* transistor utilizes either electrons or holes for current flow, and this type is described in Chapter 6.

3.2. The Bipolar Planar Transistor

A section through a planar transistor is shown in Fig. 3.1a. Here it will be seen that the *n*-type silicon substrate forms the collector region of the transistor, the collector contact being made to the back of the substrate. The first diffusion is a *p*-type, which forms the base region, and this is followed by a second, *n*-type diffusion, which forms the emitter region. The diffusion steps are carried out as described in § 2.4. The junctions of the transistor are buried in the silicon, and, as shown, the surface is coated with a protective layer of oxide. The oxide is etched, of course, to accept the metallization pattern as described in § 2.5. Because all the diffusions take place through the one plane, the transistor is termed a *planar* type, and it is basic to the fabrication of silicon integrated circuits described in § 3.4.

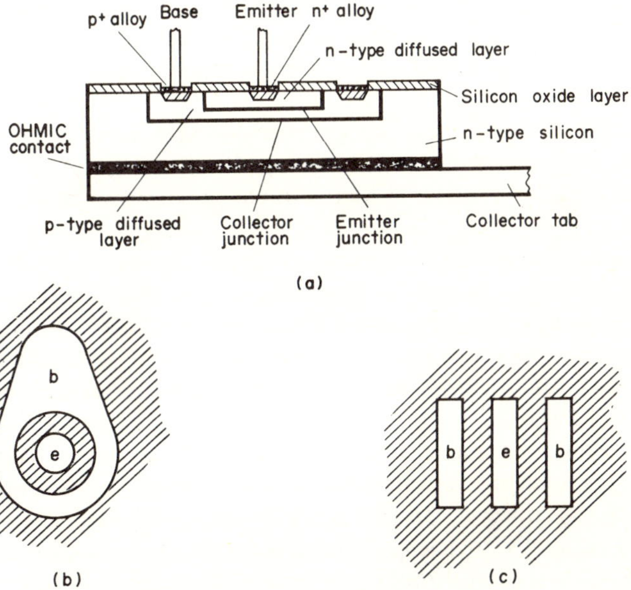

FIG. 3.1. Planar transistor geometries: (a) section through *n—p—n* transistor (Courtesy Dummer: *Electronic Components*, Pergamon Press); (b) pear-drop geometry; (c) stripe geometry.

Early designs of planar diffused transistors had circular diffused emitters and possibly an elongated base diffusion, as shown in Fig. 3.1b, the enlarged offset base area simplifying external base lead connection. However, in the diffused base, the impurity density is greatest at the surface and least at the base-collector junction, giving rise to a similar conductivity profile within the base. As a result the forward-bias voltage on the base-emitter junction is non-uniform, being greatest at the upper (surface) edge, and tailing off to little or none at the centre of the junction area. The emitter current is therefore crowded towards the emitter edge and the current handling capacity is dependent on the emitter periphery. Increasing the periphery by increasing the area is undesirable, especially at high frequencies since this also increases inter-electrode capacitance. For a given periphery P, the ratio of periphery-to-area for a circular geometry is $4\pi/P$, while for a rectangular geometry of aspect ratio a (aspect ratio is the ratio of length to width) the periphery-to-area ratio is $4(a+1)^2/a.P$, or approximately $4a/P$ for a much greater than unity. Thus the stripe geometry shown in Fig. 3.1c is preferred, for which a is much greater than unity since this gives a higher periphery-to-area ratio. Where high power at high frequencies is required, the periphery-to-area ratio is increased by means of special geometries as discussed in § 11.4.

The transistor shown in Fig. 3.1a has the disadvantage of having a comparatively wide collector region, this being necessary to provide mechanical strength to the substrate. The disadvantages of this are that it increases the collector saturation resistance (i.e. the equivalent "on" resistance of the transistor when used as a switch), and current carriers can be stored in the collector region which results in the "off" switching time being increased depending on the decay time of the stored charge.

By using an epitaxial layer the active collector region can be kept thin thus reducing the amount of stored charge. At the same time the bulk of the collector can be heavily doped to provide a low resistance collector, and is sufficiently thick to provide the necessary mechanical strength. A section through one type of epitaxial transistor is shown in Fig. 3.2a. Another form of epitaxial transistor is shown in Fig. 3.2b, known as an epitaxial base transistor. By using the epitaxial layer as the base region, the doping profile of the base can be made uniform. This means that the base–emitter breakdown voltage is not limited to the same comparatively low values (below 25 V) encountered with diffused-base units.

Figure 3.2c shows the shape of the output characteristics for the bipolar transistors discussed and it will be seen that the saturation resistance of the epitaxial transistor is an order of magnitude lower than the planar diffused unit.

(a)

(b)

(c)

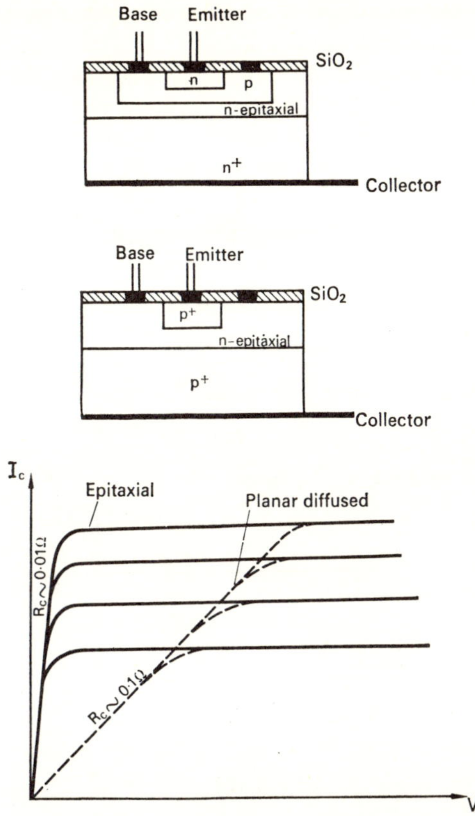

FIG. 3.2. Sketch of output characteristics showing improvement in R_c obtained with the epitaxial transistor.

3.3. The Planar Diode

Diodes are also made by the diffusion process, and a combination of diffusion and epitaxial growth. For high power applications the diodes should have large area junctions from which heat can be readily removed. Figure 3.3a shows in cross-section a typical diffused junction diode. For power applications the junction area may be of the order of 5 mm^2.

Figure 3.3b shows in cross-section a junction diode suitable for computer switching applications. Here the requirements are small junction area to maintain low capacitance (e.g. less than 2 pF), and low forward resistance. The low forward resistance is achieved using

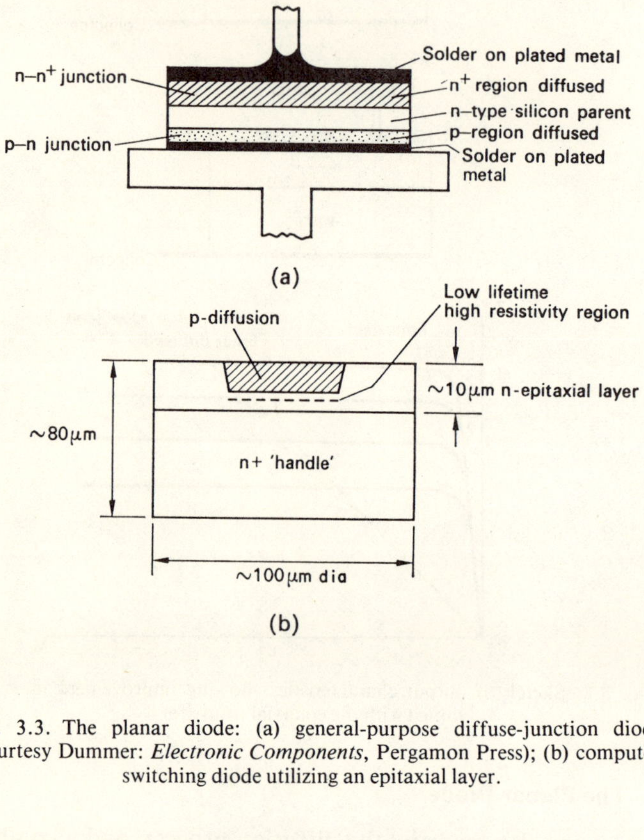

FIG. 3.3. The planar diode: (a) general-purpose diffuse-junction diode (Courtesy Dummer: *Electronic Components*, Pergamon Press); (b) computer switching diode utilizing an epitaxial layer.

an epitaxial layer grown on an n^+-substrate. The n^+-layer lowers the voltage breakdown of the diode, but fortunately for computer switching applications this does not have to be high, of the order of 10 V. As with the transistor, the thin epitaxial active region reduces carrier storage effects.

3.4. The Silicon Integrated Circuit

In the silicon integrated circuit all the components of the circuit are fabricated in the one silicon chip, and are connected together by means of a metallized layer deposited on top of the chip. Strictly speaking, this is a monolithic integrated circuit, which should be distinguished from integrated circuits in which thin-film components are deposited on top of the chip (see Chapter 10).

Transistors and diodes are comparatively easy to make and require little area, while diffused resistors, although also comparatively easy to make, require a large area, which is a disadvantage. Capacitors in monolithic form are reverse-biased diodes and as such have a number of disadvantages. They require comparatively large areas, the capacitance is voltage dependent, and the capacitors have parasitic elements associated with them; these are discussed later.

Tolerances on diffused-type resistors and capacitors are wide, and the range of values achievable is very limited. The result is that hybrid circuits, which utilize film-type resistors and capacitors are of much more practical use. Hybrid circuits are discussed in Chapter 10, while here the monolithic-type circuit is described. It should be noted that a diffused-type inductor is not a practical component, and the only solution to the inductor problem in integrated circuits is either to simulate, or eliminate, inductance by circuit design.

Resistors

The simplest form of resistor is the diffused layer to which end-contacts are made, as shown in Fig. 3.4a and b. The resistance of such a layer is given by:

$$R = \frac{\rho}{t} \cdot \frac{L}{W}, \tag{3.1}$$

where ρ is the resistivity of the material, t the thickness of the layer, L the length, and W the width. Equation (3.1) is only approximately true for the layer shown in Fig. 3.4a, since the end-contacts are not actually made to the cross-section of area tW.

When dealing with the resistance of thin sheets it is more convenient

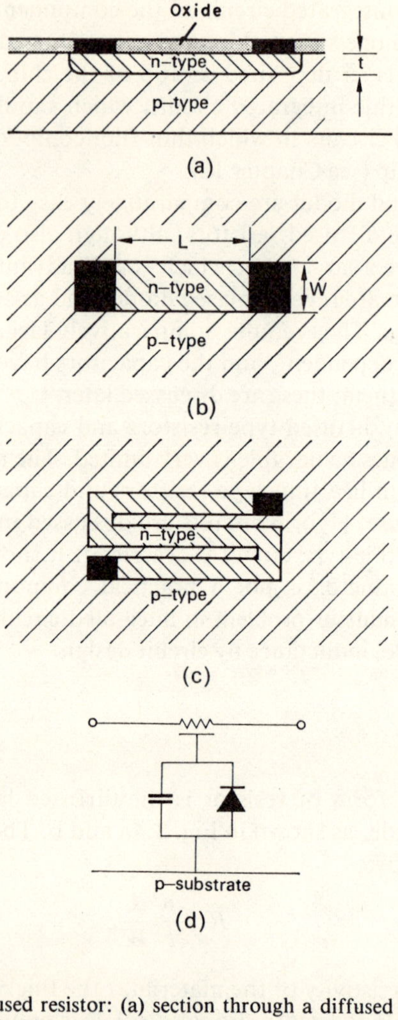

FIG. 3.4. the diffused resistor: (a) section through a diffused resistive layer; (b) plan view of (a); (c) meandered resistive layer; (d) the equivalent circuit of a diffused layer.

to work in terms of *sheet resistivity* rather than the resistivity of the material. The sheet resistivity is defined as:

$$R_s = \frac{\rho}{t} \text{ ohms per square.} \tag{3.2}$$

It is not necessary to specify the units for the square since the value of R_s is identical for any square, e.g. square inches, square metres, etc. It is essential, however, that the units be consistent. If ρ is in ohm-metres then t must be in metres and R_s will be in ohms per square; if ρ is in microhm-inches then t must be in inches, and R_s will be in microhms per square.

The ratio L/W is defined as the *aspect ratio* of the resistor and denoting this by a, then:

$$R = R_s \,.\, a. \tag{3.3}$$

For a given diffusion the sheet resistivity will be fixed and where different value resistors are required, the aspect ratio must be chosen accordingly. For high-value resistors it may be necessary to meander the path length L as shown in Fig. 3.4c in order to increase the aspect ratio.

Isolation of the resistor from other components in the chip is usually required. This is achieved by reverse-biasing the junction formed between the *n*-type resistive layer and the *p*-type substrate. The equivalent circuit for the resistor with reverse-bias on the junction is shown in Fig. 3.4d where it will be seen that the resistor has distributed capacitance to the substrate in parallel with a reverse-biased diode. The leakage current of the diode is of the order of 10 pA, and the distributed capacitance of the order of 1 pF/mm^2. (For a 1 kΩ resistor the capacitance is in the range 1–10 pF). The reverse voltage breakdown of the diode is of the order of 50 V. Circuit design must take into account these parasitic elements.

In some situations it is not practical to bias the resistor to provide isolation and a double-diffused resistor must be used as shown in Fig. 3.5a and b. An intermediate *n*-diffused region isolates the resistive layer (the top *p*-region) from the *p*-type substrate. The equivalent circuit is shown in Fig. 3.5c. In addition to the desired resistance R between A and B there exists a distributed capacitance between the *p*-region (R) and the *n*-isolation region, in shunt with the *p–n* junction formed by these two layers. The *n*-region has a further distributed capacitance to the *p*-type substrate and a *p–n* junction which forms another diode. The diodes together exhibit transistor action, and are represented in the equivalent circuit by the transistors at each side to indicate the distributed nature of the circuit. The remaining resistance in the equivalent circuit represents the resistance of the *p*-type substrate as seen from the substrate connection.

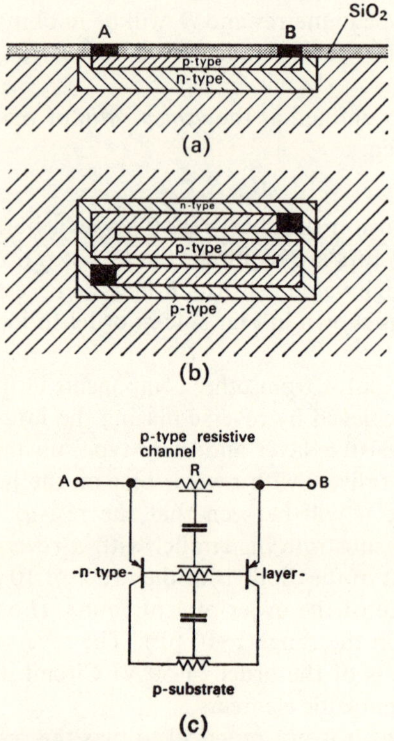

FIG. 3.5. A double-diffused resistor: (a) section through the diffused layers; (b) the plan view of (a); (c) the equivalent circuit.

The characteristics of resistors made in integrated circuit form have about ±20 per cent tolerance, 0.2 per cent per degree temperature coefficient, and range in value from 10 Ω to 100 kΩ.

Capacitors

A *p–n* junction when reverse-biased behaves as a capacitor in which the depletion layer forms the dielectric, and the *p* and *n* regions on either side the electrodes. Such capacitors are far removed from the ideal capacitor as there are a number of parasitic elements involved; also the capacitance is voltage dependent (in certain applications, e.g. varactors, this property is directly utilized). The leakage current (or reverse bias current) of the diode introduces losses which are undesirable. The maximum value of capacitance obtainable with a single junction, without going to unduly large areas, is limited to about 1000 pF. In order to increase the available capacitance using diffusion

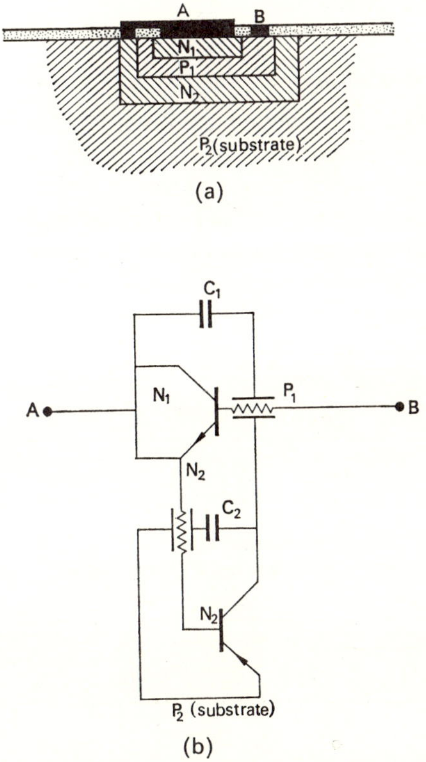

FIG. 3.6. The double-diffused capacitor: (a) section through the diffused layers; (b) the equivalent circuit.

methods a double-diffusion-type capacitor is usually employed, as shown in Fig. 3.6a. The two *n*-type diffusion regions are connected together to form one electrode, and the *p*-type region sandwiched between these forms the other electrode. The correct bias must be maintained across the capacitor, i.e. the *n*-type electrode is biased positive with respect to the *p*-type. The equivalent circuit is shown in Fig. 3.6b. The desired capacitance is C_1 in parallel with C_2 between the terminals *A* and *B*. The capacitance is accompanied by the distributed resistance of the layers along with the *p–n* junctions which form the two transistors having the configuration shown in Fig. 3.6b. The maximum capacitance that can be achieved using the sandwich type structure is about 2000 pF/mm^2.

Isolation and Buried n$^+$*-layers*

Isolation by means of a reverse-biased *p–n* junction has already been described in connection with resistors and capacitors. A similar technique is used to isolate regions in the layer in which a complete circuit is to be made; the regions so isolated are termed *lands*, a separate component being built into each land. Figure 3.7 shows how isolation is achieved. Assuming that an *n*-type epitaxial layer has been grown and which also contains n^+-regions (these will be discussed shortly) as

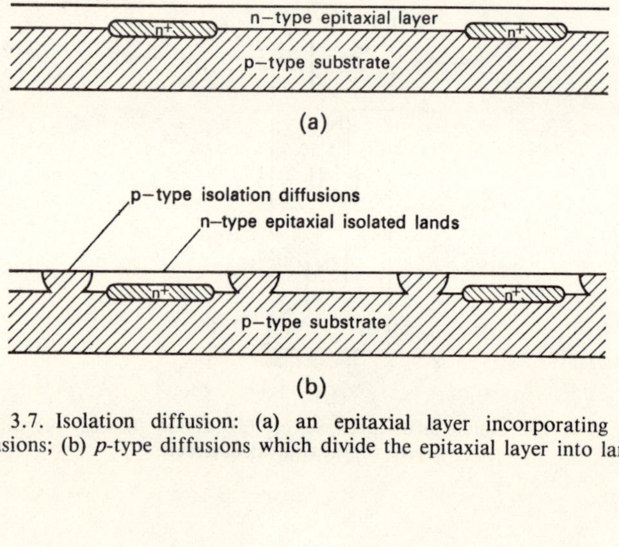

FIG. 3.7. Isolation diffusion: (a) an epitaxial layer incorporating n^+-diffusions; (b) *p*-type diffusions which divide the epitaxial layer into lands.

shown in Fig. 3.7a, then *p*-type diffusions are made through the epitaxial layer to meet up with the *p*-type substrate, as shown in Fig. 3.7b. In this way *n*-type lands are formed and to complete the isolation these must be biased positively with respect to the substrate.

When a transistor is formed in a land the *n*-type epitaxial layer forms the collector, the connection to which must be made through the top surface. In order, therefore, to reduce the effective series resistance of the collector path, the n^+-region is diffused in before the epitaxial layer is grown. In the subsequent growth of the epitaxial layer the n^+-region diffuses upwards with the epitaxial layer and therefore the lands intended for transistors have the final profile as shown in Fig. 3.7b.

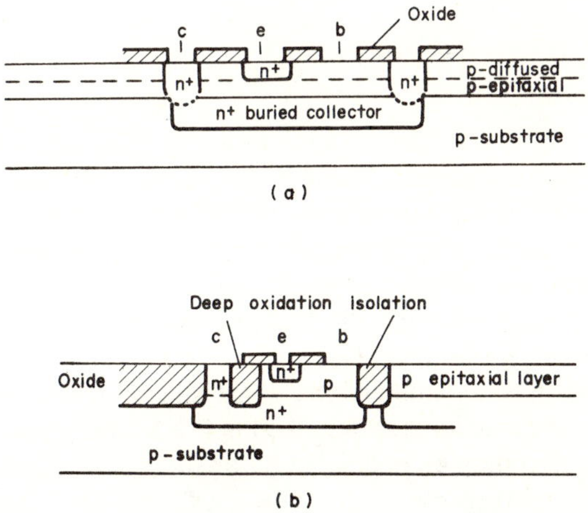

FIG. 3.8. (a) Collector diffusion isolation; (b) isoplanar isolation.

Isolation as described above is extragavant of silicon area and more recent methods have been aimed at reducing the overall area required. Two of these which are of significance are the *collector diffusion isolation* (CDI) first described by the Bell Telephone Labs, and the Isoplanar method, by the Fairchild company.

In the CDI method, n^+ regions are formed in the *p*-type substrate as before, but now a *p*-type epitaxial base region is grown. Although not essential, a *p*-type diffusion is also made into the *p*-epitaxial layer as it gives better control of surface properties, and of the sheet resistivity, which is important for fabrication of resistors in the same layer. An n^+ isolation diffusion is then carried round the buried n^+ collector as shown in Fig. 3.8a so that a *p*-epitaxial land is effectively formed within the collector. An *n*-type emitter may be diffused in and contacts formed in the normal way.

In the Isoplanar method, the lands are isolated by means of deep oxidation, Fig. 3.8b. In implementing the method use is made of silicon nitride as a masking agent as this can be etched independently of the oxide (selective etching) and growth of the oxide has negligible effect on the nitride.

Using these methods area reductions of the order of 50 per cent or greater can be achieved.

Interconnections

In Fig. 3.9 is shown a diffused resistor and a transistor which form part of an integrated circuit. The emitter of the transistor is shown connected to the resistor by means of an aluminium strip. In a complete integrated circuit the components are interconnected by means of an interconnection pattern deposited over the top of the silicon chip. The conductor material is usually aluminium, and it is deposited by means of vacuum evaporation (see Chapter 8.) To obtain the desired pattern, further masking and etching of the silicon oxide, and the aluminium, must be carried out as described in § 2.5. More than one interconnection pattern may be deposited by including an insulating layer (usually SiO_2) between conductor layers except where feed-through connections are required. Although multi-layer circuits are possible, more than two conductor patterns are seldom encountered in practice.

In Fig. 3.10 is shown an example of a complete silicon integrated circuit. The actual circuit is shown in Fig. 3.10a, and a plan view of the chip in Fig. 3.10b. The white strips indicate the aluminium interconnections, and the numbers shown correspond with the connection numbers shown on the circuit of Fig. 3.10a. The chip size is 1.5 mm × 1.6 mm. Compared to a similar circuit constructed out of discrete components, the silicon integrated circuit is seen to be a remarkable technological achievement, and yet the circuit shown in Fig. 3.10 is classed as small scale integration. Present day semiconductor memories (see Chapter 12) achieve packing densities of some thousands of circuit elements on single chips typically 3 mm × 4 mm, this being an example of large scale integration (LSI).

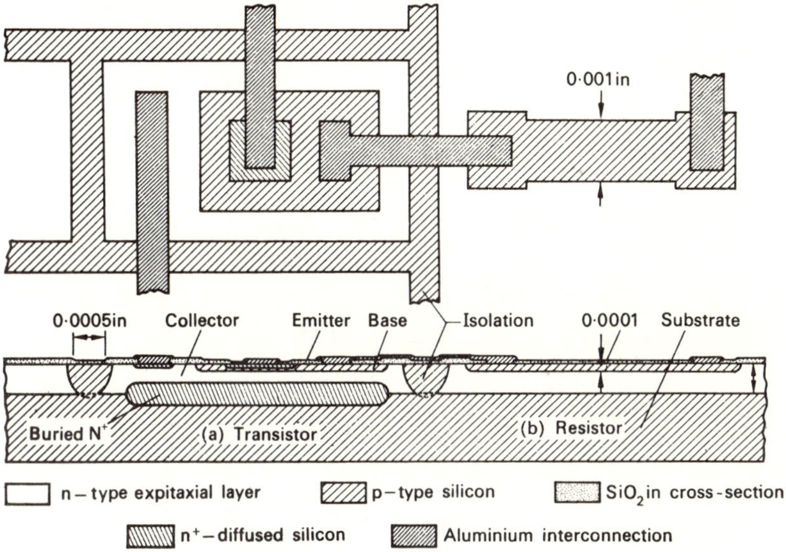

FIG. 3.9. Interconnection of two integrated circuit components. (Courtesy: I.E.E. and Electronics & Power, and the Plessey Co. Ltd.)

Figure 3.10c shows the silicon slice, 1¼ in. diameter, which contains some hundreds of the circuit chips. At this stage the individual circuits are tested, usually by means of a simple d.c. test, and faulty chips marked. This enables the faulty chips to be rejected before the bonding and packaging stages which follow as these are very costly in comparison with the rest of the processing. The silicon slice is scribed along fine lines running between the individual circuits and broken into the individual chips.

Packaging

To complete the assembly the silicon chip must be mounted in a suitable holder which allows leads to be taken to the chip, and which also provides protection against mechanical damage and against adverse ambient conditions. This final stage is termed *packaging*. Figure 3.11 shows two commonly used types of packs (the top covers have been removed for clarity) and it will be seen that the bulk of the completed circuits is the pack.

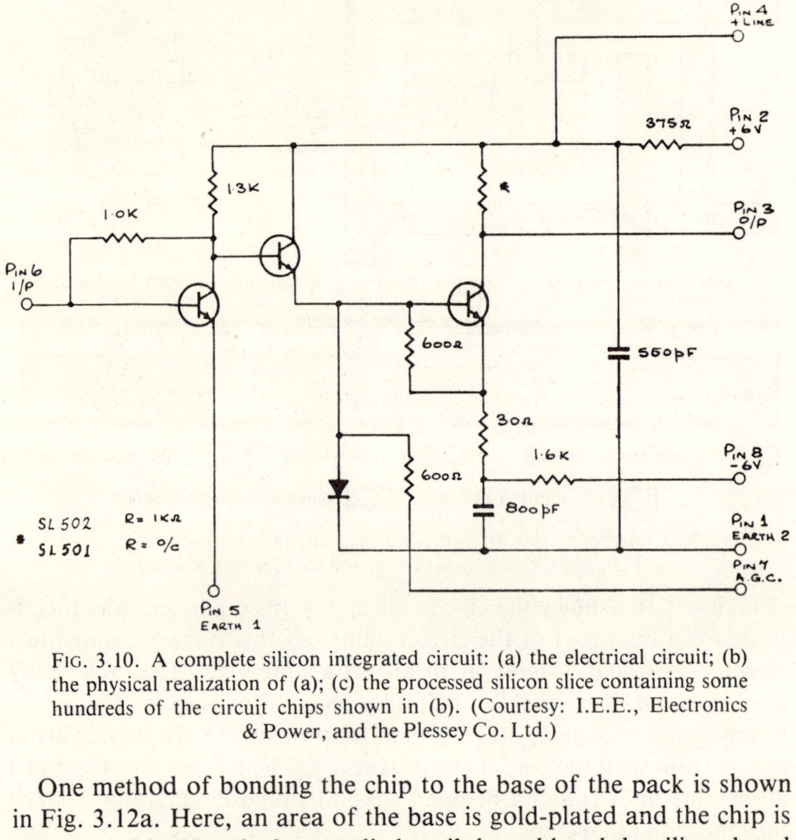

FIG. 3.10. A complete silicon integrated circuit: (a) the electrical circuit; (b) the physical realization of (a); (c) the processed silicon slice containing some hundreds of the circuit chips shown in (b). (Courtesy: I.E.E., Electronics & Power, and the Plessey Co. Ltd.)

One method of bonding the chip to the base of the pack is shown in Fig. 3.12a. Here, an area of the base is gold-plated and the chip is placed on this. Heat is then applied until the gold and the silicon bond together, in what is known as a eutectic bond.

Connections between the chip and the external connections on the pack are usually made by means of wire bonds, as shown in Fig. 3.12b, fine wire being used (e.g. 0.001 in. diameter aluminium). The wire

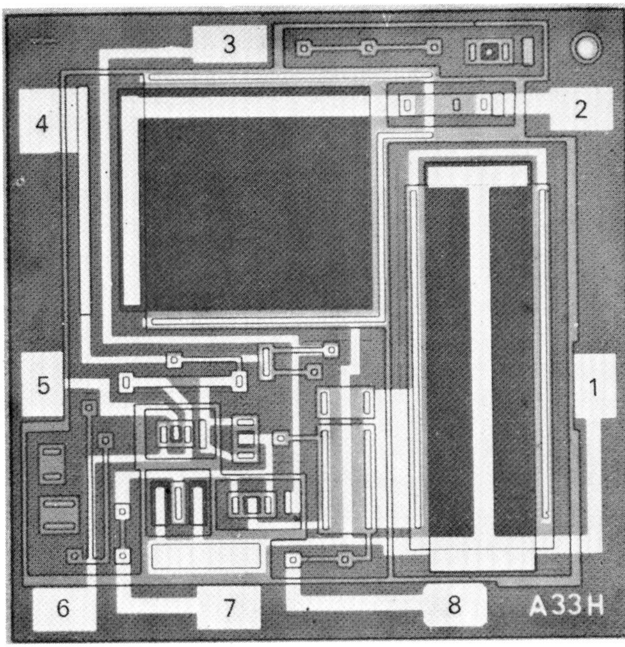

FIG. 3.10 (b).

FIG. 3.10 (c).

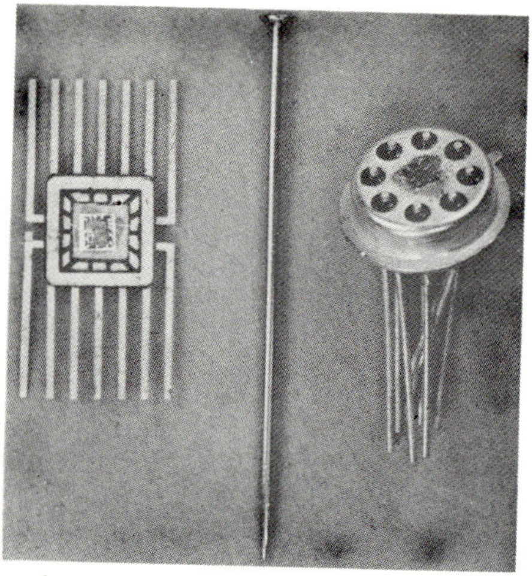

FIG. 3.11. Integrated circuit packaging: left, the flat-pack; right, the TO–5 can; 1 in. pin shown for scale. (Courtesy: I.E.E., Electronics & Power, and the Plessey Co. Ltd.)

may either be bonded using thermal compression bonding, or using ultrasonic bonding. Both methods are described in Chapter 10.

Some form of cover has to be fitted to complete the package. With metal packs the lid may be soldered, or cold-welded on, and it is necessary to make a hermetic seal. More recently, plastic encapsulation has been used for packaging. Although this is attractive, as it is cheap and easy to apply, it does not provide as good a hermetic seal as metal, and metal-ceramic packages. However, it is more than adequate for most applications in the domestic equipment field.

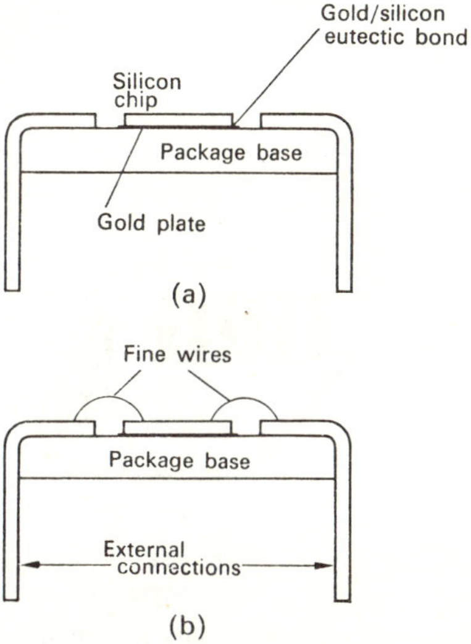

FIG. 3.12. (a) Silicon chip assembled to package by means of a gold–silicon eutectic alloy bond. (b) Connections from chip circuit to external pins made by means of fine wire bonds.

Ceramic packages are widely used for LSI circuits, the most popular type being the dual in-line package (DIP). The main body of the package is usually made up in three layers, the top layer bearing a sealing ring to which a metal cover can subsequently be attached, the centre layer carrying the interconnections necessary between the external pins (lead frame) and the chip, and the lower layer, to which the chip is bonded. A typical assembled DIP is shown in Fig. 3.13, the ceramic version being shown in Fig. 3.13a, and the plastic version in Fig. 3.13b.

Note the metal lid on the ceramic type. For certain types of semiconductor memories the lid may house a window as shown in Fig. 12.8c.

Where the LSI circuit requires more than about 40 external leads, ceramic type packages are generally used, because they are easier to produce to the rather stringent specifications compared with plastics, and they provide better hermetic sealing, and better heat dissipation.

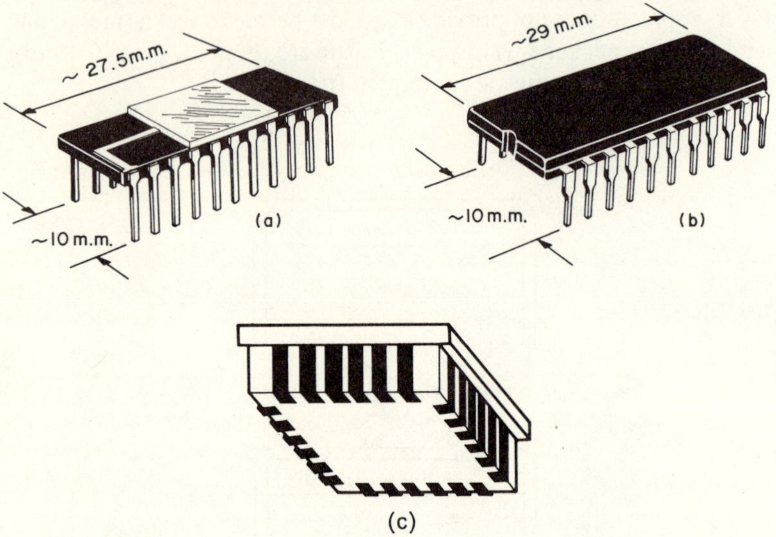

FIG. 3.13. LSI packages: (a) ceramic D.I.P.; (b) plastic D.I.P.; (c) ceramic carrier.

Various other types of packages are in use for LSI circuits. Chip carriers for example, are small ceramic (or glass-ceramic) carriers usually assembled from three layers similar to the DIP, but with the external leads being carried down the side as shown in Fig. 3.13c. These can be connected into printed circuit boards using the reflow solder method (see § 10.3).

In the manufacturing process for ceramic packages and carriers, the metallized alumina layers, initially uncured, have to be fired in a furnace to bind them together in the form of the final stable ceramic material. The natural colour of this is white, and black colouring material may be added where photosensitive effects are to be reduced to a minimum.

CHAPTER 4

Bipolar Logic Circuits

4.1. Introduction

One of the characteristics of digital computers, and of digital techniques generally, is the use that is made of large numbers of electronic circuits to perform certain logic functions. However, only a few basic circuit types are needed, and it is the repetitive use of given circuit types that creates the requirement for large numbers of circuits. It is precisely because integrated circuits are manufactured most economically when the requirement is for large numbers of a few basic circuit types that they have found widespread use in digital-type circuitry. In fact, ideas, which have been dormant for many years (e.g. pulse-code modulation for communications systems) because of the sheer cost and complexity of building reliable digital circuitry out of discrete components, are now becoming practical as a result of integrated circuits. It is also an interesting thought that the use of integrated circuits in computer design has made available, at realistic cost, computers which can be used to aid in the design of integrated circuits!

In this chapter, some of the basic logic circuits available in integrated circuit form will be described in order to illustrate this major field of application. Only the bipolar type of circuit (see § 3.1) will be described here, as the unipolar type devices are covered in Chapter 7.

A logic circuit is one which performs functions according to certain rules termed *logic rules.* For example, if a circuit has three inputs and is required to produce an output voltage when all three inputs are at a given voltage level, the circuit is known as an AND logic circuit (or AND *gate*). In effect, all three inputs, *A and B and C*, must be at the required level for an output to result.

With *binary logic* a circuit need only be able to distinguish between two specified voltage levels (e.g. voltage off—voltage on) and these are referred to as logic levels $\underline{0}$ and $\underline{1}$. The logic operation of the circuit may be described in terms of these levels, it being unnecessary to know the actual voltage levels involved, but when it is desired to relate the logic levels to the circuit conditions a clear statement must be made regarding the representation of logic levels through voltage polarity. *Negative logic* means that logic level 0 is represented by zero (or ground) potential, and logic level 1 by a negative voltage at the point, with respect to ground. This definition can, in fact, be widened by saying that logic level 0 is always less negative than logic level 1. *Positive logic* is when logic level 1 is represented by a positive voltage at the point, and logic level 0 by zero voltage; more generally, logic level 1 is always more positive than logic level 0.

4.2. Basic Logic Functions

The logic functions which are most frequently used are:

(i) The AND function, in which a logic level 1 is required at all inputs simultaneously to produce a logic level 1 at the output.
(ii) The OR function, in which a logic level 1 at any input will produce a logic level 1 at the output.
(iii) The NOT function, in which a logic level 1 input produces a logic level 0 output, and a logic level 0 input produces a logic level 1 output. (Also known as a NEGATE function.)
(iv) The NAND function, which is an AND function and NOT function combined, the NOT function following the AND function.
(v) The NOR function, which is an OR function combined with a NOT function, the NOT function following the OR function.

It can be shown that all logic functions can be performed by various combinations of NOR gates alone (logic circuits are commonly known as *gates* because they can either let a logic signal through, or block it). In practice it is easier to have gates which perform the specific functions listed above. The logic symbols for the basic gates are shown in Fig. 4.1, along with the logic equations which describe the function. As shown, the NOT function is also referred to as an INVERTER.

Note that a circuit which performs an AND function in positive logic automatically performs an OR function in negative logic, and vice-versa.

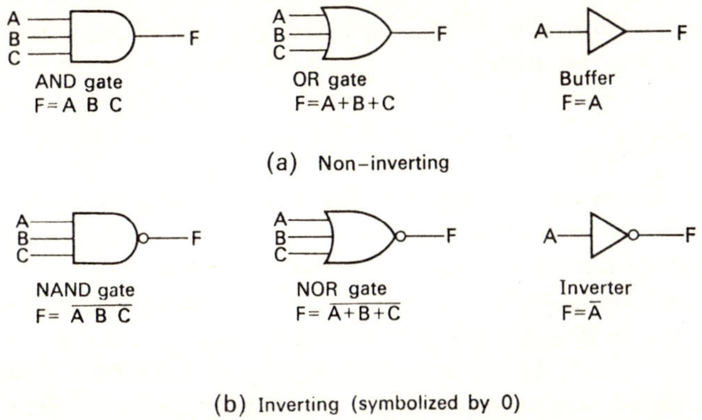

FIG. 4.1. Logic symbols for basic gate functions. (Courtesy: Motorola Inc.)

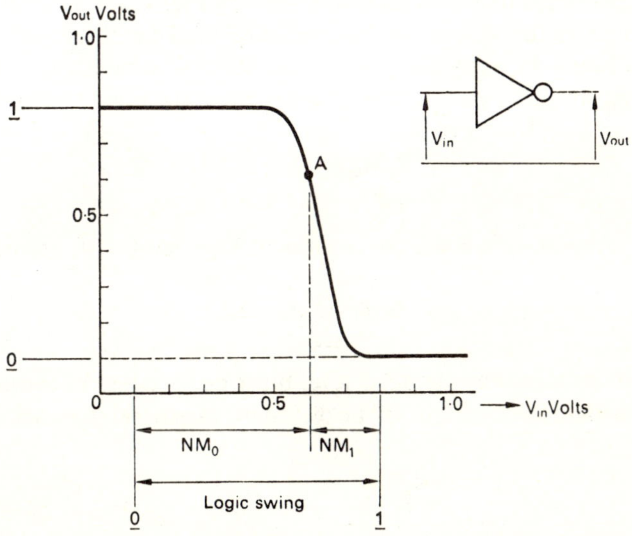

FIG. 4.2. The voltage transfer characteristic, used to illustrate noise margin.

4.3. Performance Specifications

The main factors which have to be considered when assessing a circuit's performance are: (i) noise margin, (ii) speed, (iii) loading effects, (iv) temperature effects, (v) power consumption.

Noise Margin

This is a measure of a gate's ability to avoid being switched on by a noise pulse. Noise, in this context, is an electrical disturbance or signal at the input other than the desired signal. Noise may be induced into a gate circuit from the switching action of a neighbouring circuit or it may be externally generated noise picked up on the input lead. In either case it is obviously desirable that the gate circuit should not operate on a noise signal. The *noise margin* is defined in relation to the voltage transfer characteristic of the circuit, shown in Fig. 4.2. For clarity, the characteristic for a single-input NOR (i.e. inverter) stage is shown.

When the input voltage is at logic level $\underline{0}$ (0.1 V) the output voltage is at logic level $\underline{1}$ (0.8 V). If the input level is now altered to logic level $\underline{1}$, the output drops to logic level $\underline{0}$, as shown by Fig. 4.2. The noise margin is defined as the voltage difference, measured on the input voltage axis, between the unity gap point A and the operating point. Denoting the noise margin by NM, then the noise margin when the input is at logic level $\underline{0}$, is from Fig. 4.2:

$$\begin{aligned} NM_0 &= 0.6 - 0.1 \\ &= 0.5 \text{ V.} \end{aligned} \tag{4.1}$$

The noise margin when the input is at logic level $\underline{1}$ is likewise given by:

$$\begin{aligned} NM_1 &= 0.8 - 0.6 \\ &= 0.2 \text{ V.} \end{aligned} \tag{4.2}$$

It will be seen that the noise level must be specified in conjunction with the logic level since it will be different, in general, for each.

Speed

The speed of a logic circuit refers to how fast the output level can follow changes in the input level. Ideally the output level should follow instantly any change at the input; in practice a number of factors contribute to a delay occurring between input and output. At the input, the driver resistance (usually the collector saturation resistance of the previous stage) in conjunction with the input capacitance of the gate results in a delay proportional to the *R–C* time constant. Likewise, at the output, the output resistance of the gate in conjunction with the input capacitance of the following gates (there may be more than one) contributes a delay proportional to the output time-constant. Since this delay depends on the output loading, the loading must be stated when specifying speed.

Within the gate, the transit time for the current carriers across the base region, and carrier density build-up time, will add to the delay when the circuit is being switched "on". The time taken for the removal of stored charge mainly from the collector region will add to the delay for the "off" switching. As a result, the "on" delay time will not, in general, be equal to the "off" delay time, and the average delay time of these two is used to specify the gate delay.

Delay times are measured from the 50 per cent levels as shown in Fig. 4.3, the delay time being specified as:

$$\bar{t}_d = \tfrac{1}{2}(t_{d1} + t_{d2}). \tag{4.3}$$

Note that the length of the pulse T does not enter into the expression for delay time.

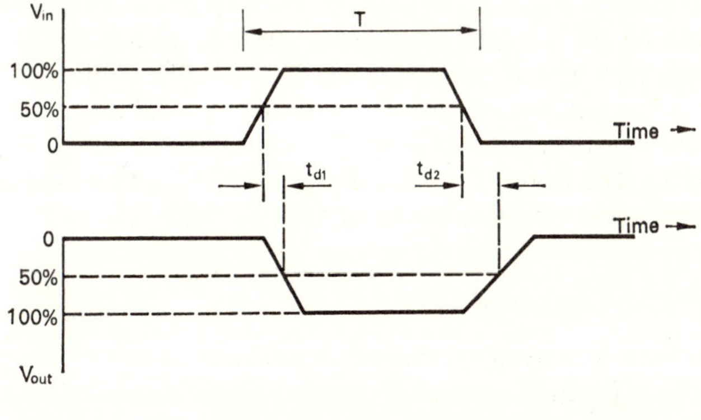

FIG. 4.3. ON delay time t_{d1} and OFF delay time t_{d2}.

Loading Effects (Fan-in and Fan-out)

Fan-out refers to the number of gates that can be operated in parallel by the gate to which the specification applies. Fan-in refers to the number of inputs that a gate can handle. Fan-in and fan-out greatly affect operating parameters such as noise margin and speed, and the limit to the degradation of these also sets the limit to the loading. In addition, fan-out is limited in some circuits by the way in which the output current divides between the gates connected to the output; where one gate takes a disproportionate share of the current the problem is termed *current hogging*, (see TTL).

Temperature Effects

Changes in temperature will affect parameters such as transistor gains, resistor tolerances, and this will obviously be reflected in changes in performance specifications. For normal commercial applications circuits are designed to operate over the range 0 to 75°C. For more rigorous applications such as military use or space communications applications, the temperature range may be —55 to 125°C but such increase in performance is only achieved at greater cost.

Power Consumption

The amount of power dissipated in a gate is important as it determines the temperature rise of the gate. Where a large number of gates are incorporated on a single substrate or printed circuit board the power dissipation may well be the limiting factor on the number of circuits that can be used. For purposes of design the power dissipation is defined as the power dissipated with the gate operating on a 50 per cent duty-cycle. The Power × Delay Time product is often used as a figure of merit, some typical valves being listed in Table 7.1.

Saturated and Non-saturated Logic

When the base current to a transistor is sufficient to move the operating point to a position such as *A* on the output characteristics as shown in Fig. 4.4, the transistor is said to be in saturation. Many logic circuits are saturated when in the "on" state, and this has the advantage of providing a highly stable "on" condition. The main disadvantage of saturated logic is that carrier storage in the collector region is increased which reduces the speed. In saturation the collector–base junction becomes forward biased which results in minority carrier diffusion in both directions across the collector–base junction. When the transistor is switched off time must be allowed for the minority carrier densities to decay and usually the decay time of those in the collector region dominates. In the *p–n–p* transistor electrons are the minority carriers in the *p*-type collector, and this is gold-doped to speed up the recombination of electrons with holes as mentioned in § 2.4.1. With non-saturated logic the circuit design ensures that the transistors do not enter the saturation condition and as a result very much faster operation can be achieved. This is obtained at the expense of an increase in power consumption.

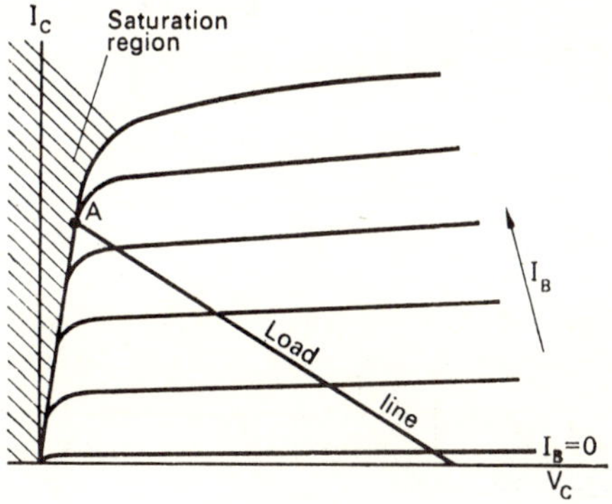

FIG. 4.4. The saturation region of a transistor; the operating point in saturation is *A*.

4.4. Logic Circuits

Diode Transistor Logic (DTL)

In Fig. 4.5a is shown a DTL circuit which performs a NAND function. The circuit is shown using positive logic i.e. logic, level 1 is positive with respect to ground, while logic level 0 is at approximately

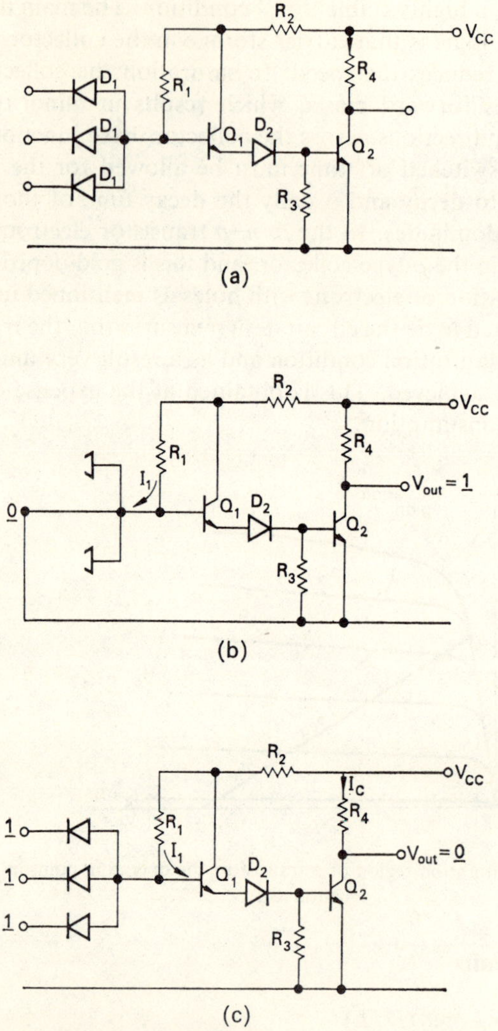

FIG. 4.5. Diode transistor logic (DTL): (a) the actual circuit; (b) the equivalent circuit with any one input at logic level 0; (c) the equivalent circuit with all inputs at logic level 1.

ground potential. When the input to any one of the diodes D_1 is at logic level 0 the current through R_1 is shunted through that diode, away from the transistor Q_1. Figure 4.5b shows this condition for the centre diode at logic level 0. To ensure that the current chooses the D_1 path the diode D_2 is included in the Q_1 transistor emitter path. D_2 is known as a *voltage offset diode* and is always forward-biased. Since the transistor input has effectively two diodes in series, i.e. the base–emitter diode of the transistor Q_1 in series with D_2, the current will choose the diode D_1 path when this is forward-biased. Under these conditions transistor Q_1 is biased to cut-off, and therefore no current flows into the base of Q_2. The output voltage of Q_2 therefore rises to the positive level V_{CC} which is logic level 1.

When all the inputs are raised to logic level 1 as in Fig. 4.5c all of the input diodes will be reverse-biased and therefore the current I_1 will flow into the base of the transistor Q_1. Transistor Q_1 now conducts which in turn forward-biases transistor Q_2 causing it to conduct. As a result the output voltage drops to $V_{CC}—I_cR_4$, where I_c is the "on" current of Q_2. This sets the logic zero level.

It will be seen, therefore, that when all inputs are at logic level 1, the output is at logic level 0, which is the NAND function.

Transistor Q_1 does not saturate since the base current is fed from the collector voltage through R_1. The inverter transistor Q_2 does saturate and hence the speed of the DTL circuit is not particularly fast. The circuit produces a large logic swing and has a good noise margin. It is widely used for medium-speed applications.

Transistor–transistor Logic (TTL or T^2L)

The TTL circuit is similar to the DTL circuit but it utilizes a feature unique to integrated circuit technology, the multiple emitter transistor. The basic circuit for TTL is shown in Fig. 4.6a. The multiple emitter transistor is made using the standard diffusion processes described in Chapter 3, it only being necessary to etch the required number of emitter windows in the oxide for emitter diffusion. Three emitters are shown in Fig. 4.6a although the number can be increased to about ten if required. The area for the multiple emitter transistor is not greatly increased over that needed for a single emitter transistor.

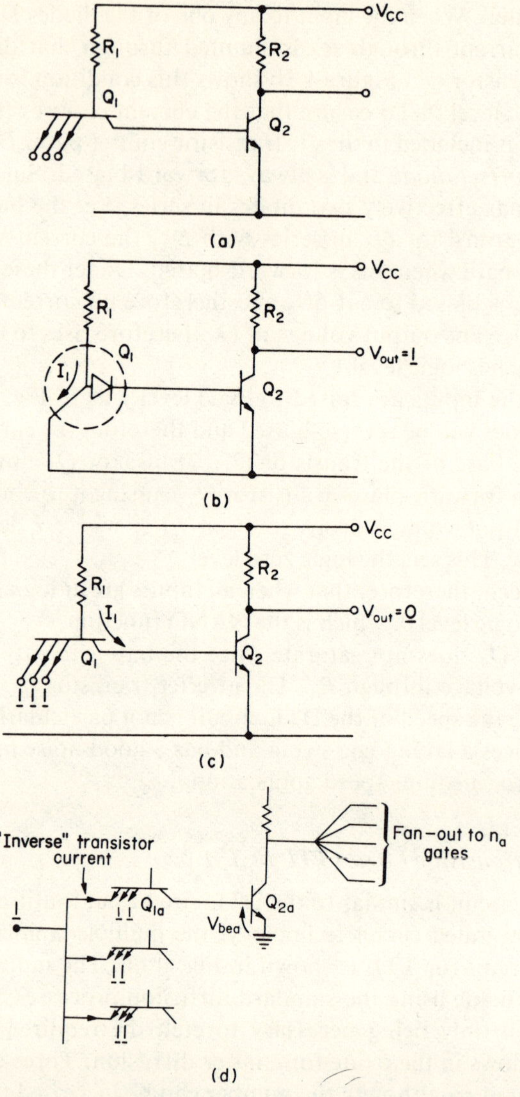

FIG. 4.6. Transistor transistor logic (TTL): (a) the basic circuit; (b) the equivalent circuit with any one input at logic level 0; (c) the equivalent circuit with all inputs at logic level 1; (d) current hogging conditions.

The multiple base-emitter diodes correspond to the input diodes of the DTL circuit, and the base–collector diode of transistor Q_1 obviates the need for a voltage offset diode (D_2 in the DTL circuit). It is interesting to note that under static conditions (i.e. when the input and output levels are constant) the base–collector junction of Q_1 is always forward-biased.

When any one of the inputs is at logic level $\underline{0}$ the base current flows through that emitter and hence is shunted away from the input of Q_2. The equivalent circuit for this condition is shown in Fig. 4.6b, where the forward-biased input emitter is shown as a short circuit to ground, and the forward-biased base–collector junction is represented by a diode. This junction has to develop forward-bias conditions for circuit equilibrium, since reverse bias would mean Q_1 operating in the normal active mode which in turn would require a steady state reverse base current from Q_2, which cannot happen. The voltage appearing across the emitter-base diode of Q_2 is insufficient to turn Q_2 on since the *total* voltage appearing across this diode in series with the equivalent Q_1 base–collector diode is the voltage drop across a single diode, the emitter-base diode of Q_1.

Because Q_2 is cut-off its output voltage rises to V_{CC} or logic level $\underline{1}$. If now, all inputs are taken to logic level $\underline{1}$ the input junctions are reverse-biased and the base current to Q_1 flows through the forward-biased base–collector diode and into the base of Q_2, switching this on. The output then drops to logic level $\underline{0}$ as indicated in Fig. 4.6c.

TTL is saturated logic but during the period of the switching-off transient the collector–base junction of Q_1 becomes reverse-biased and Q_1 works as a normal transistor, thus drawing a reverse base current from Q_2. This speeds up considerably the withdrawal of stored charge from Q_2 and as a result TTL is faster than DTL.

One of the problems with TTL is current hogging, where, because of variations in device parameters, current may be shunted away from one or more of the inputs into the other input circuits. This arises when all the inputs are at logic level $\underline{1}$. Q_1 is then in the *reverse active mode*; the current gain is normally very small in this mode but none-the-less, a current flows from input to the base of Q_2 (in addition to the base–collector current of Q_1), as a result of the inverse transistor action of Q_1. A number of gates are shown operating in parallel, Fig.

4.6d, with all the inputs at logic level 1 (for clarity only one input is shown for each gate). The base–emitter voltage of Q_2 in saturation (for each Q_2 gate) increases with increase in Q_2 collector current, which in turn increases with fan-out on Q_2. For example, V_{bea} depends on fan-out n_a, Fig. 4.6d. Increased fan-out can result in V_{be} increasing sufficiently for the Q_1 inverse transistor current to be diverted at the input i.e. it is "hogged" by the other gates, and Q_2 for the particular gate that has excessive fan-out comes out of saturation, with a resulting switching error.

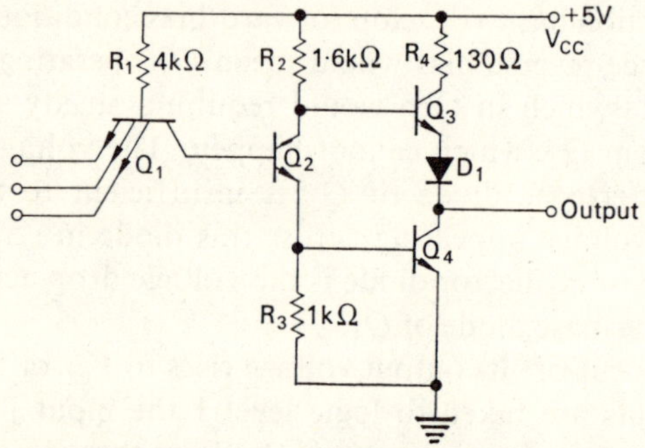

FIG. 4.7. TTL incorporating a totem-pole output circuit. (Courtesy: Electronic Components, and Texas Instruments, Bedford.)

In some versions of TTL an improved output drive capability is achieved by using a push–pull arrangement of transistors, as shown in Fig. 4.7. The output circuit (which is, of course, part of the complete integrated circuit) consists of transistors Q_3 and Q_4, diode D_1 and resistor R_4 in series. This is known as a *totem-pole* arrangement. When the output of Q_2 is at logic level 0 transistor Q_3 is cut-off while Q_4 conducts because part of the emitter current from Q_2 flows into the base of Q_4. The output is pulled down to logic level 0 as Q_4 is driven into saturation. The diode D_1 is a voltage offset diode which ensures that Q_3 cuts off when Q_2 is saturated, even at high temperatures. Resistor R_4 is a current-limiting resistor.

When the output of Q_2 is at logic level 1 transistor Q_3 conducts but

Q_4 desaturates through R_3 and cuts off. As a result the output rises to logic level 1. The totem-pole arrangement provides a greater swing in logic level than would otherwise be obtainable, but it will be noted that in the steady state "on" or "off" condition the totem-pole branch does not draw current and so there is very little additional power dissipation.

Emitter Coupled Logic (ECL)

The basic circuit for emitter-coupled logic is shown in Fig. 4.8a. It is possible to obtain either a NOR function or an OR function, depending on which output terminal is selected. The logic levels are defined with reference to the base voltage V_4 of transistor Q_4, logic level 1 being more positive than V_4 while logic level 0 is less positive. Consider first the situation when all the inputs to transistors Q_1, Q_2, and Q_3 are at logic level 0. These transistors will be biased near the cut-off and the total circuit current, which is determined mostly by R_2, will flow through transistor Q_4. By selecting resistor R_1 such that the voltage drop across it when carrying the total circuit current is equal to the logic swing V_L (the logic voltage swing is the difference between the two logic levels), the output voltage from the collector of Q_4 will be $V_{CC} - V_L$ for the above conditions. The output voltage at the other terminal is V_{CC}. Denoting logic level 1 voltage by V_1 and logic level 0 voltage by V_0 the output voltage from Q_4 may be written as $(V_{CC} - V_1) + V_0$ (i.e. V_L is replaced by $V_1 - V_0$). The output voltage from the other output terminal V_{CC} may be written as $(V_{CC} - V_1) + V_1$. The reason for writing the output voltages in this way is to show they each represent a given logic level shifted by amount $(V_{CC} - V_1)$. This shift in level can be eliminated as described below.

If now, any one of the transistors Q_1, Q_2 or Q_3, has its input lifted to logic level 1 it will be forward-biased to a greater extent than Q_4 and therefore the current will be shunted from Q_4 to the forward-biased transistor. This results in a switch of output voltage levels between the two output terminals. Thus the output from Q_4 gives the OR function while that from the common node gives the NOR function, in both cases shifted by amount $(V_{CC} - V_1)$.

In order to eliminate the voltage shift of level ($V_{CC}-V_1$) the outputs are taken through emitter follower circuits as shown in Fig. 4.8b, referred to as d.c. translators. The levels are adjusted so that ($V_{CC}-V_1$) is exactly offset by the base-emitter voltage of the emitter followers, this making the output levels equal to the logic levels. Input and output levels are required to be compatible in order that multi-gate systems may easily be implemented. The emitter follower also reduces the output impedance of the gate which increases the fan-out capability. By suitable choice of operating levels and use of the d.c. translator,

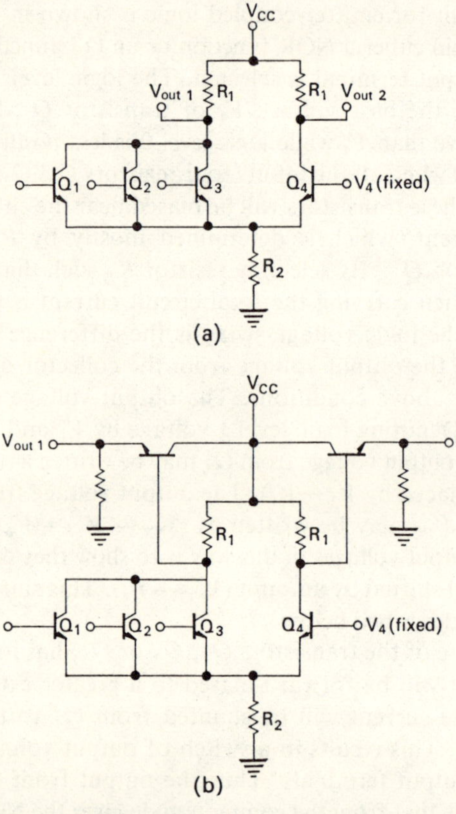

FIG. 4.8. Emitter-coupled logic (ECL): (a) the basic circuit; (b) the circuit incorporating emitter follower outputs.

emitter coupled logic can be made compatible with other logic families. An alternative name for emitter coupled logic is *current mode logic* (CML) this arising because a constant current is switched from one branch to another in the gate.

The circuit design for ECL is such that no transistor operates in saturation, therefore the circuit is very fast. This increase in speed compared with other types of logic circuits is gained at the expense of greater power consumption.

4.5. Comparison of Logic Gates

DTL is usually classed as medium speed logic, its delay time being of the order of 30 nsec. It has a good noise margin and low power consumption. For most applications not requiring high speed it is probably the most satisfactory.

TTL is high speed, a typical delay time being of the order of 8 nsec. The current hogging problem tends to limit fan-out and internally generated noise can be high. These defects can be minimized by suitable design, and as a result TTL finds wide application for high-speed applications.

ECL is the fastest of the three, a typical delay time being 5 nsec, but its power consumption is high compared with other types of logic. The noise margin also tends to be worse compared with other types but to offset this, this circuit itself generates less noise because it is non-saturated logic.

TABLE 4.1.

COMPARISON OF CHIP SIZES

J–K Flip-flop type	Chip size, mils (1 mil = 0.001 in.)
DTL	48×48
TTL	72×75
ECL	58×65

Regarding the size of chip, this will depend on the total circuit function. Data for a special type of logic circuit, known as a J–K Flip-flop, is given in Table 4.1. (Data supplied courtesy of Motorola Inc.)

It will be seen that the DTL chip is significantly smaller than the other two. Although the ECL chip is smaller than the TTL chip the TTL design incorporates more test devices for this particular size, which allows better control to be achieved of process variations and circuit parameters. It should be noted that the TTL circuit is very dependent on transistor gain (beta) while the ECL circuit is very dependent on resistor ratio with less dependence on transistor beta.

Taking all factors into account, DTL is cheapest and TTL dearest. It should be noted that other forms of logic circuits are available and space has only permitted the three most representative types to be selected for comparison. Other types are described briefly below and in chapter 7.

4.6. Schottky TTL

As an alternative to gold doping to improve switching speed, a Schottky diode may be formed across the collector–base junction of Q_2, which acts as a clamping diode to prevent Q_2 from going into saturation. The structure is illustrated in Fig. 4.9a, and the equivalent circuit in Fig. 4.9b. Referring to Fig. 4.9a, aluminium acts as a *p*-type impurity in silicon, so that the aluminium to *p*-type base forms an ohmic contact, whereas the aluminium to *n*-type silicon collector forms a rectifying contact as described in § 1.6. For this particular combination the turn-on voltage of the diode is about 0.3V, compared to about 0.6 V for the on-voltage of a *p–n* junction in silicon; thus the Schottky diode clamps the collector–base diode of the transistor at 0.3 V, preventing this from entering saturation. Effectively, the TTL is converted to a non-saturating logic with consequent improvement in speed. The advantages over gold doping are that the production yield is better, and the same transistor type can be used for linear applications.

4.7. Merged Transistor Logic

This type of logic is also known as Integrated Current Logic, or I^2L. Fig. 4.10a shows the basic NOR circuit. Current I_B is supplied to the base of each switching transistor. Either of the inputs A or B going to logic level $\underline{1}$ steers I_B into the base of the corresponding transistor which is then switched on, pulling the collector voltage down to logic

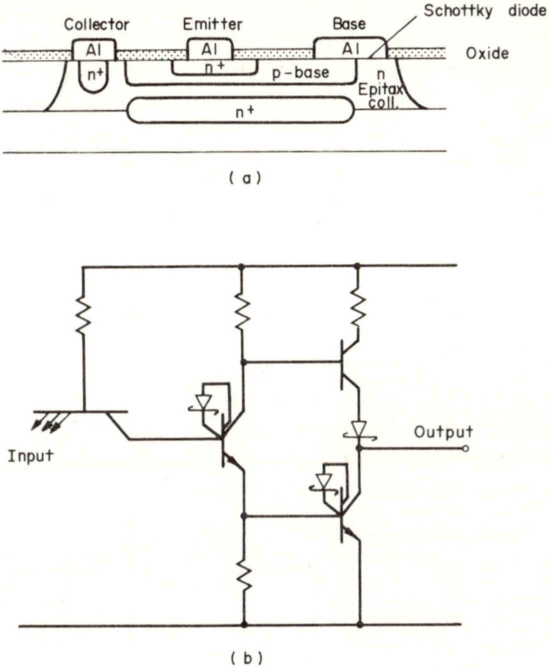

FIG. 4.9. Schottky TTL: (a) structure; (b) equivalent circuit.

level 0. Figure 4.10b shows how the current supply is provided in Merged Transistor Logic (MTL). The external supply I is connected to the p_1 emitter of the p_1–n_1–p_2 transistor. A single transistor only, of this kind is needed in the physical structure although this is shown as two equivalent units in the equivalent circuit. Two switching transistors are illustrated, a single collector n_1–p_2–n_2 to the left of the current supply transistor, and a dual collector unit n_1–p_2–n_2 to the right. The n_1 layer is seen to be common to the emitters of the switching transistors and the base of the current supply transistor, and hence the name *merged* transistor logic. Minority carriers injected from emitter p_1 into base n_1 are collected by the p_2 collectors where they automatically become the injected base currents for the n_1–p_2–n_2 switching transistors.

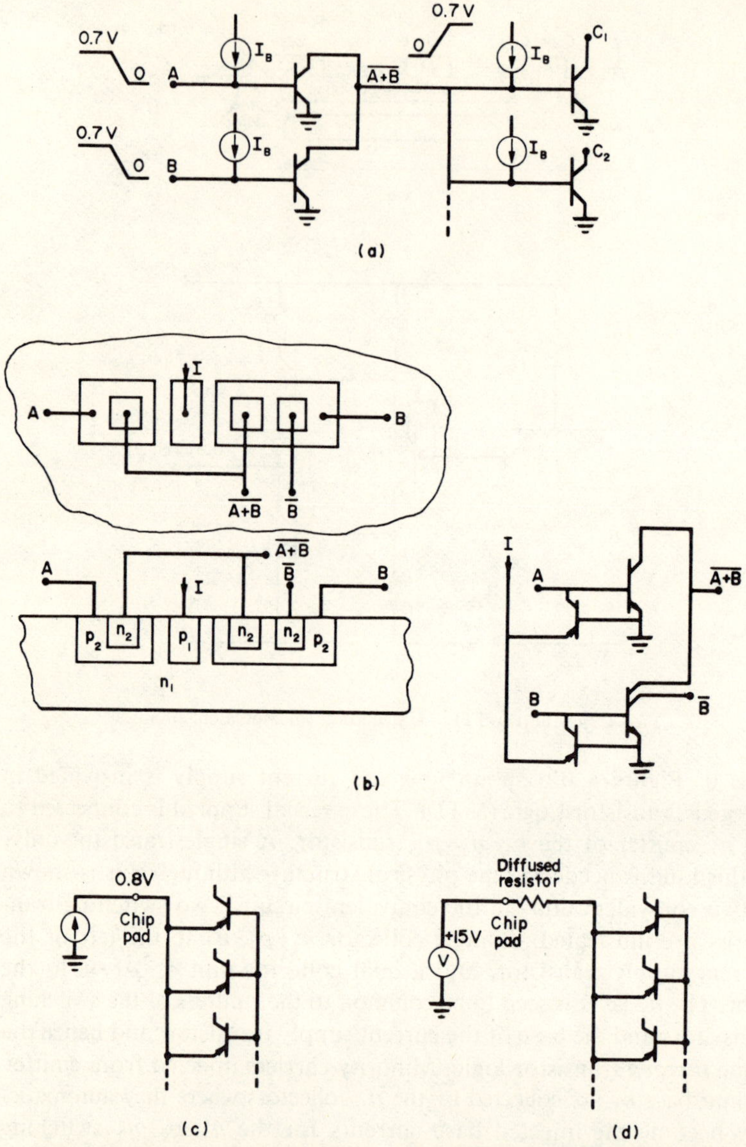

FIG. 4.10. Merged transistor logic: (a) basic NOR gate; (b) structure and equivalent circuit; (c) constant current supply; (d) constant voltage supply.

The external supply may be a constant current source as shown in Fig. 4.10c, which results in about 0.8 V at the chip supply pad. Alternatively, a +15 V supply may be connected through a resistor as shown in Fig. 4.10d. Here, the resistor is fabricated as an *n*-diffused component at the same time as the collectors of the *n–p–n* transistors.

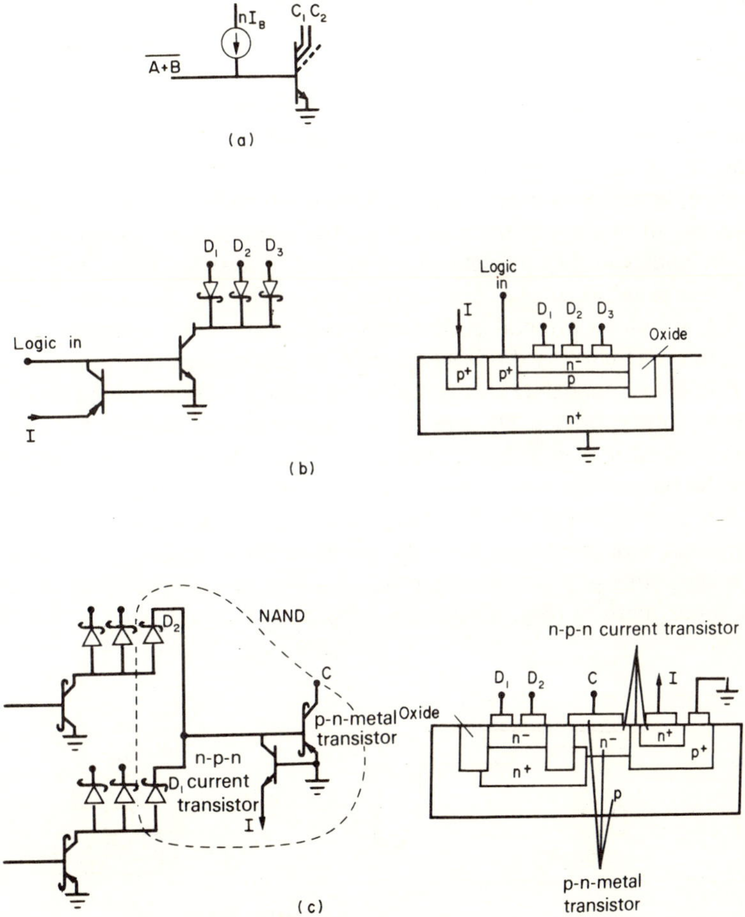

FIG. 4.11. Developments from merged transistor logic: (a) multicollector unit; (b) Schottky coupled transistor logic; (c) Schottky transistor logic.

As a further step in the merged structure, where the output A + B is fed to succeeding inverters, as shown in Fig. 4.10a, it is seen that the bases are common, and this leads to a multicollector unit the equivalent circuit for which is shown in Fig. 4.11a. C_1 and C_2 of this correspond to C_1 and C_2 in Fig. 4.10a. Fabrication of the multicollector structure is similar to that of the dual collector unit shown in Fig. 4.10b.

A disadvantage of the multicollector unit is that the transistor gain has to be high in order that the collectors can act independently of one another, and the high gain results in high charge storage which increases delay time (as already discussed for saturated logic). By utilizing Schottky diodes instead of multicollectors a lower gain transistor can be used, as the diodes decouple the outputs from one another. A possible circuit arrangement, along with the structure, is shown in Fig. 4.11b, and is known as Schottky Coupled Transistor Logic (SCTL). This has been taken one step further in the Schottky Transistor Logic (STL), in which a metal collector is used so that the collector–base junction is a Schottky diode which provides clamping, as discussed for TTL in § 4.6. A NAND gate using this approach is shown in Fig. 4.11c. The Schottky diodes D_1 and D_2 cannot now be merged into the metal collector to which they are connected, and are instead merged into the base of the following gate. Here the current supply transistor is an *n–p–n* type and the switching transistor a *p–n*-metal, the reason being that with present technology it is easier to make Schottky diodes on *n*-type silicon. Note that deep oxide isolation similar to that of the Isoplanar method (Fig. 3.8b), is used in the structures of Fig. 4.11b and c.

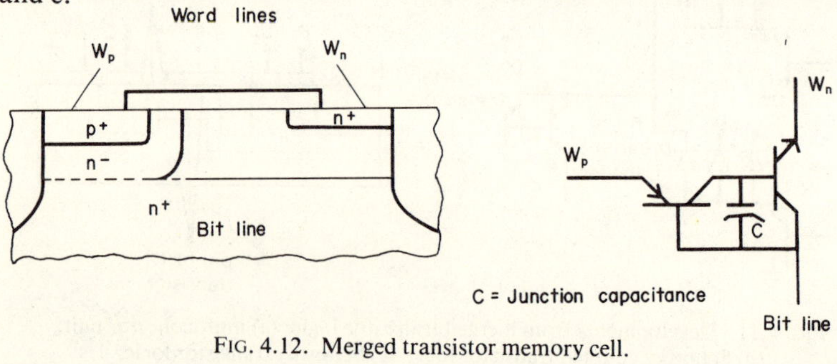

FIG. 4.12. Merged transistor memory cell.

Merged transistor structures are also used in computer memory cells employing dynamic logic as discussed in § 12.4. Figure 4.12 shows the cell structure along with the equivalent circuit. Interelectrode capacitor *C* is utilized in the memory action as described in § 12.4. Note that in this case it is the collector of one transistor that is common with the base of the other transistor.

CHAPTER 5

Differential Amplifiers

5.1. Introduction

In addition to digital circuits, operational amplifier circuits provide another area of application for integrated circuits which is growing steadily. The operational amplifier is one of the most useful and versatile circuits which has come into use in electronics, as the characteristics are determined almost solely by passive feedback components added externally to the basic amplifier. There are certain major advantages to be gained by constructing the basic amplifier as an integrated circuit (compared with a discrete component circuit). Therefore, the basic amplifier circuit, which is invariably a differential-type amplifier, will be used to illustrate this particular field of application of integrated circuits.

5.2. The Differential Amplifier

The basic circuit for a differential amplifier is shown in Fig. 5.1a. This consists of two matched sections which ideally should be identical. It is easier to approach this ideal with integrated circuit design than with discrete component design.

The input signal v_{in} is applied between the base terminals this being known as the *differential mode* (*DM*) input, since, for the polarity shown in Fig. 5.1a:

$$v_{in} = (v_1 - v_2). \qquad (5.1)$$

The output signal is given by a similar expression:

$$v_0 = (v_{01} - v_{02}). \qquad (5.2)$$

Denoting the voltage gains of the respective stages as A_1 and A_2, then eqn. (5.2) can be written as:

$$v_0 = (A_1 v_1 - A_2 v_2)$$
$$= A_1 \left(v_1 - \frac{A_2}{A_1} . v_2 \right) . \qquad (5.3)$$

When the voltage gains A_1 and A_2 are equal, eqn. (5.3) reduces to:

$$v_0 = A . v_{in}, \qquad (5.4)$$

where $A = A_1 = A_2$.

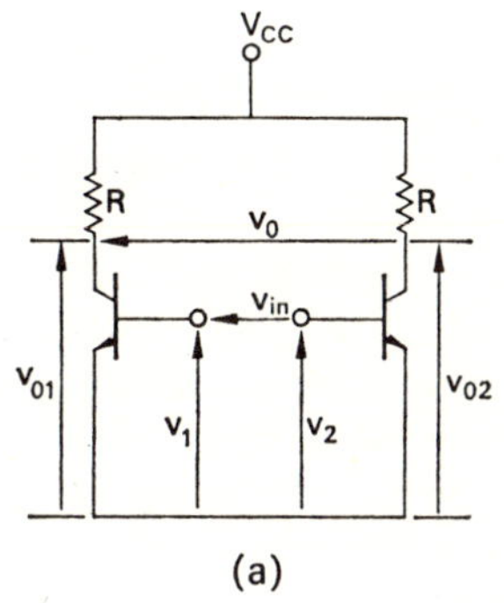

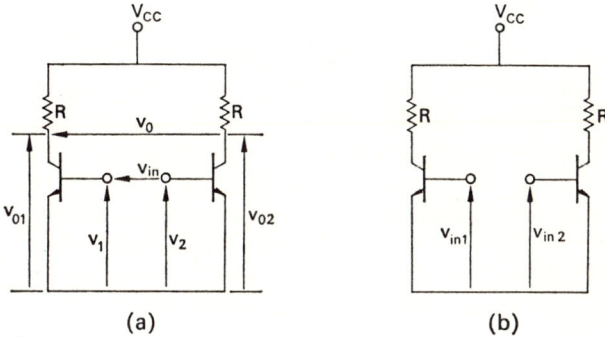

FIG. 5.1. The basic differential amplifier circuit: (a) the differential-mode (DM) input; (b) the common-mode (CM) input.

It is also possible for the differential amplifier to pick up what are usually unwanted signals at each input as shown in Fig. 5.1b. This is termed the common mode (CM) input. The CM output signal is given by eqn. (5.3) on replacing v_1 with v_{in1}, and v_2 with v_{in2}. Thus, even where v_{in1} is equal to v_{in2}, an output will still be produced when A_1 does not equal A_2. Ideally the CM output should be zero but variations between amplifier sections due to changes in temperature and mismatch introduced through manufacturing tolerances can result in an output being present. The most sensitive parameter is the base–emitter voltage (this is the offset voltage which has already been encountered in the DTL circuit of Chapter 4). For silicon the temperature coefficient for the offset voltage is 2.3 mV/°C so that if a temperature difference of 0.01°C exists between the two transistors, the differential offset voltage is 23 μV. With integrated circuits it is easier to maintain the transistors at the same temperature (compared with discrete component circuits) since they can be made very close together in the silicon chip. The drift of offset voltage with temperature for integrated circuits is reduced by a factor of about 2 : 1 compared with discrete component circuits.

The offset voltage also depends on the dimensions of the base–emitter junction. For example, if the base widths differ by 10 per cent between transistors, the differential offset voltage will be 2.5 mV. Again, with integrated circuits it is possible to match the dimensions very closely since both transistors may be made together from the same diffusions. For integrated circuits the differential offset voltage can be reduced by a factor of about 5 : 1 compared with the best matched discrete transistors.

It is desirable to have the differential amplifier fed from a constant current source as this reduces the cross-coupling between common mode input and differential mode output, due to unbalance between the circuits. Thus, suppose a common mode signal was able to increase both transistor and collector currents by equal amounts then any unbalance between the load resistors would result in a differential mode output. A constant current source, however, would not permit any increase in current at all, and therefore no output would result. With a desired differential-mode input, as the collector current in one-half increased, that in the other half would decrease, thus maintaining a constant total current without affecting the desired differential mode operation.

An approximation to a constant current source is achieved by feeding the emitters through a large resistor R_E (of the order of 10 kΩ) (Fig. 5.2a), but obviously a limit is set on R_E by the available voltage supply V_{CC}. Ideally, if R_E could be made infinite cross-coupling between common-mode input and differential-mode output would be zero. A better approach to a constant current source can be achieved by using a transistor biased for constant current. In fact, with integrated circuits it is easier to make a transistor than a large value resistor for R_E. A suitable circuit is shown in Fig. 5.2b. Here, the constant current source is obtained through transistor Q_1, which will have an output resistance of the order of 1 mΩ. The current through the diode and resistor chain R_2 and R_1 is considerably greater than that through Q_1, with the result that the bias voltage to the base of Q_1 is determined mainly by the diode. The diode also provides temperature compensation as it tracks with the base–emitter diode of Q_1.

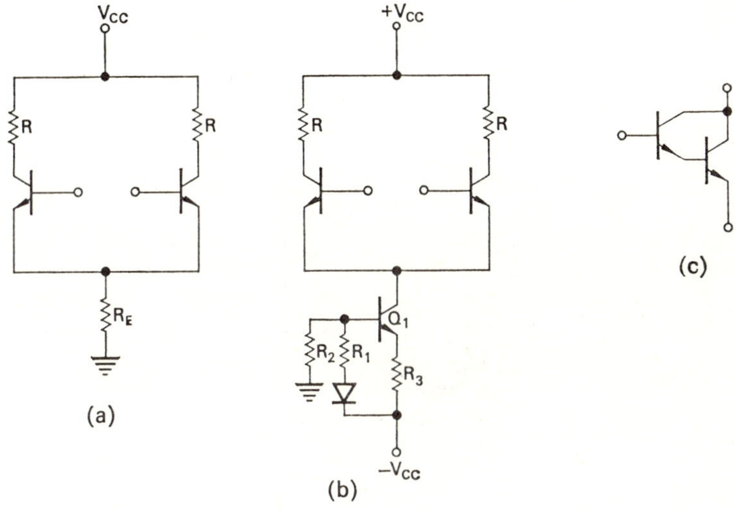

FIG. 5.2. (a) A constant current source approximation using large R_E; (b) a better approximation, using a transistor; (c) a Darlington-pair connection.

In practice, a complete amplifier will usually consist of a number of differential stages with their constant current sources, and possibly incorporating the compound transistor arrangement known as the Darlington pair. The basic circuit for this is shown in Fig. 5.2c, and the main properties of the Darlington circuit are very high input impedance and current gain, and very low output impedance when used as an emitter follower circuit. The complete amplifier will also usually include a single-ended output circuit.

It is not possible here to go into the design detail of these amplifiers, but Fig. 5.3 shows the complexity which can easily be achieved in integrated circuit form. The unit shown consists of *two* identical amplifiers, *each* amplifier consisting of two differential stages followed by single-ended output stages as shown in the circuit of Fig. 5.3a. Each amplifier can be used independently of the other, and the simplified circuit representation is shown in Fig. 5.3b. The use of this type of equivalent circuit is discussed further in the following section.

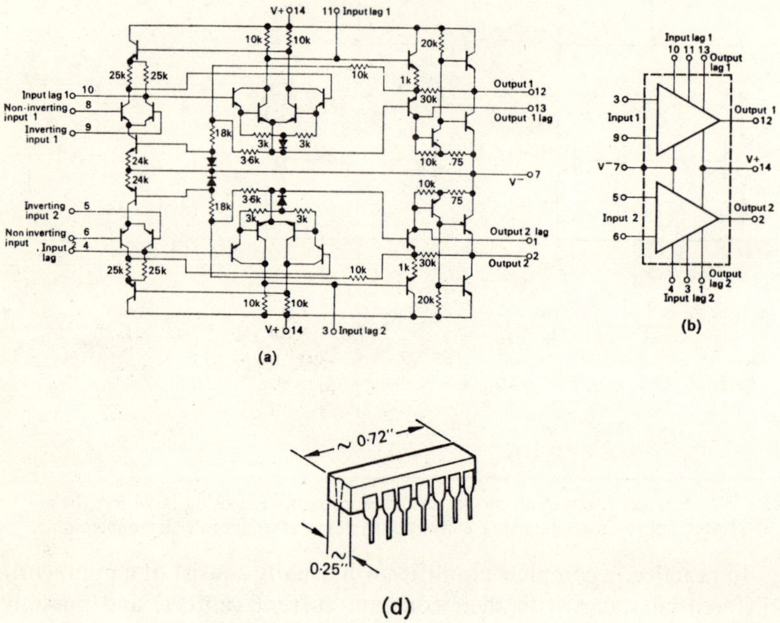

FIG. 5.3. A dual-integrated operational amplifier: (a) the circuit; (b) the equivalent circuit; (c) see Plate facing; (d) the packaged unit. (Courtesy: Motorola Inc.)

Figure 5.3c shows a microphotograph of the silicon chip containing both amplifiers (where the connection pads have been numbered for convenience in relating the layout to the circuit of Fig. 5.3a). This vividly illustrates the component density achievable with integrated circuit methods. Figure 5.3d shows the dual amplifier in its package.

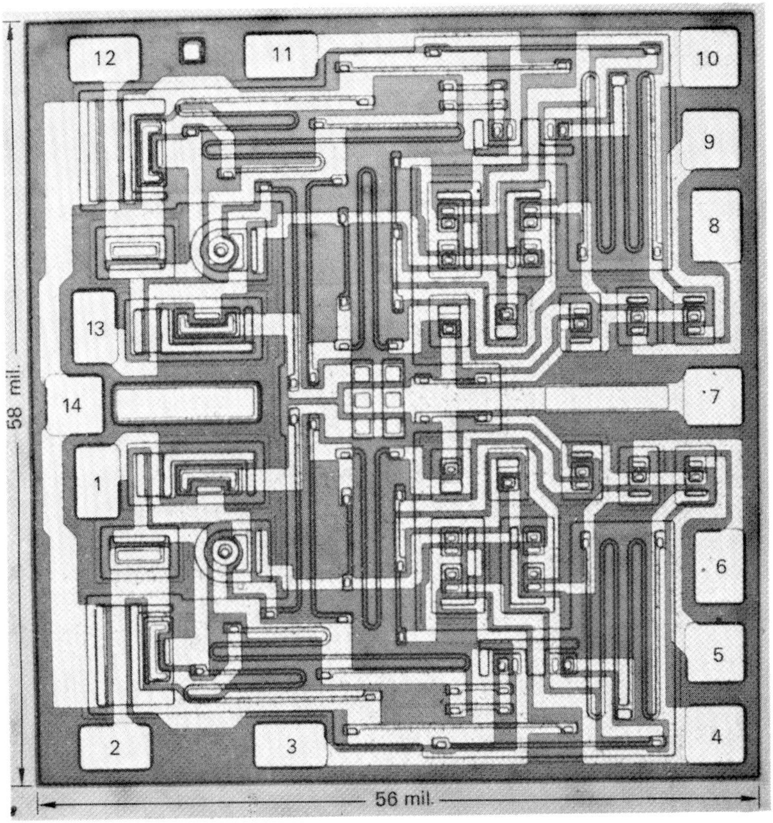

FIG 5.3. (c) A photomicrograph of the silicon chip containing the dual amplifier.

5.3. The Integrated Operational Amplifier

The operational amplifier was so named originally because it was used to perform certain mathematical operations such as addition and integration. It was soon realized that these amplifiers could be used in a wide variety of applications, especially in instrumentation and control engineering. The basis of the operational amplifier is the differential amplifier.

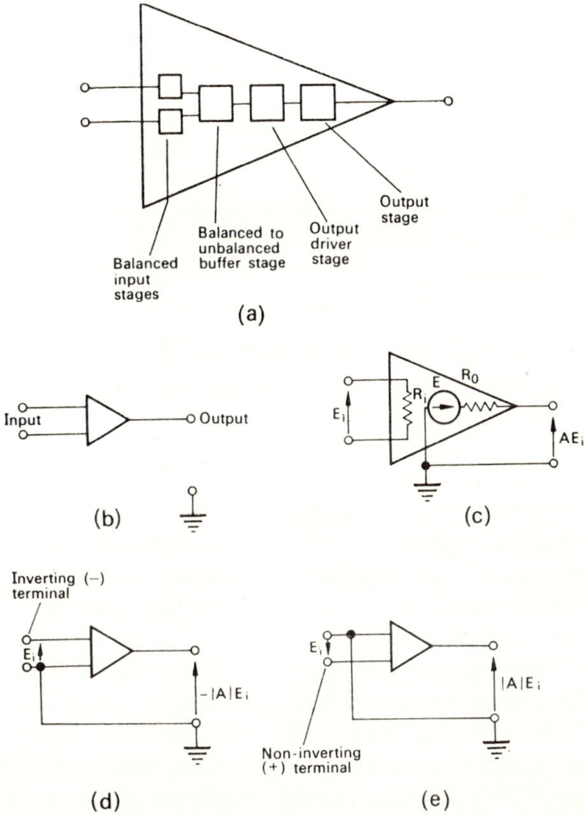

FIG. 5.4. Operational amplifier representation for open-loop conditions: (a) block diagram of internal stages; (b) a simplified representation of (a); (c) defining the terminal parameters of the amplifier; (d) the inverting connection; (e) the non-inverting connection.

From the point of view of application the operational amplifier may be represented as shown in Fig. 5.4a which may be further reduced to the form shown in Fig. 5.4b. It is important to know the terminal characteristics of the amplifier. These characteristics may be represented on a circuit as shown in Fig. 5.4c, and they are summarized in Table 5.1.

TABLE 5.1.
OPEN-LOOP CHARACTERISTICS FOR AN OPERATIONAL AMPLIFIER

Parameter	Typical value	Ideal value
Input resistance R_i	150 kΩ	∞
Output resistance R_o	4 kΩ	0
Voltage gain A	50×10^3	∞
Gain × bandwidth product	1 MHz	∞

The characteristics given in Table 5.1 are the *open-loop* values, that is the amplifier characteristics without feedback. In use, external circuitry is used to provide a very large amount of feedback such that the overall amplifier characteristics depend on the feedback network and are independent of the open-loop characteristics. The behaviour of the amplifier including the feedback network is described in terms of the *closed-loop* characteristics. It is this property of being able to determine the overall amplifier performance mainly by the feedback network that makes the operational amplifier so useful. It should be noted that the reason why a large feedback factor is possible is because the open-loop characteristics approach their ideal values.

In addition to a balanced input connection, inputs may be applied either as shown in Fig. 5.4d or Fig. 5.4e. With the circuit as shown in Fig. 5.4d, a 180° phase shift occurs between input voltage and output voltage therefore the input terminal free from ground is known as the *inverting terminal.* With the connection as shown in Fig. 5.4e, the output voltage is in-phase with the input voltage and therefore the input terminal above ground in this case is known as the *non-inverting input.*

Figure 5.5a shows how the operational amplifier may be used as an inverting amplifier. Assuming very high internal gain A (the ideal value) then since $V_i = V_0/A$, V_i may be assumed to be virtually zero. Assuming that the input resistance has its ideal value then the current I flows through R_1 and R_2. With V_i zero it is seen that:

$$\frac{V_1}{R_1} = -\frac{V_0}{R_2},$$

or,

$$A_v = \frac{V_0}{V_1} = -\frac{R_2}{R_1}, \qquad (5.5)$$

where A_v is the external, or closed-loop voltage gain for the circuit of Fig. 5.5a. This is seen to be determined entirely by the feedback resistors R_1 and R_2. If typical values had been used for the open-loop characteristics instead of the ideal values the closed-loop voltage gain would be found to be within 1 per cent of the value given by eqn. (5.5).

Figure 5.5b shows the circuit for a non-inverting amplifier for which the voltage gain is found to be:

$$A_v = \frac{R_1 + R_2}{R_2} \qquad (5.6)$$

Figure 5.5c shows the voltage follower circuit, the properties of which are that the output voltage V_0 is equal to the input voltage V_1, the input resistance approaches infinity, and the output resistance approaches zero. Thus, the circuit behaves as an ideal impedance transformer.

Figure 5.6 shows a simple voltage summing amplifier. This is similar to the inverting amplifier of Fig. 5.5a except that the total current I is given by:

$$I = I_1 + I_2 + I_3 + \ldots I_n, \qquad (5.7)$$

hence

$$-\frac{V_0}{R} = \frac{V_1}{R} + \frac{V_2}{R} + \frac{V_3}{R} + \ldots \frac{V_n}{R},$$

or

$$-V_0 = V_1 + V_2 + V_3 + \ldots V_n. \qquad (5.8)$$

Many other operational circuits can be devised using the basic amplifier with external feedback. As Fig. 5.3a shows, access is provided to points in the internal amplifier so that its characteristics, in

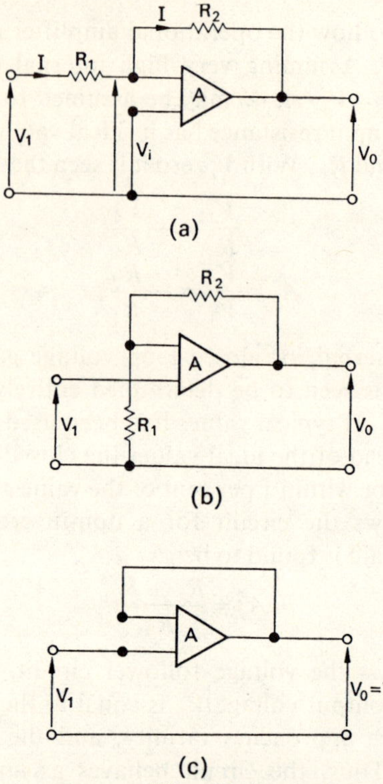

FIG. 5.5. (a) The inverting amplifier; (b) the non-inverting amplifier; (c) a voltage follower.

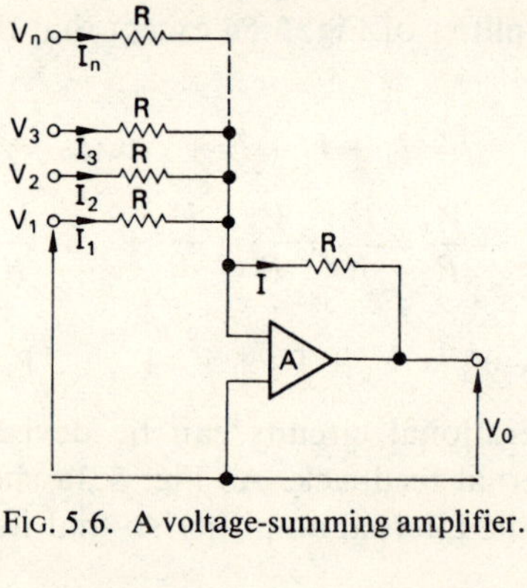

FIG. 5.6. A voltage-summing amplifier.

TABLE 5.2.

TEST CONDITIONS RELATING TO FIG. 5.7.

Figure no.	Curve no.	Voltage gain	Test conditions R_1 (Ω)	R_2 (Ω)	R_3 (Ω)	C_1 (pF)	C_2 (pF)	Output noise (mV rms)
5.7b	1	1	10 k	10 k	1.5 k	5.0 k	200	0.10
	2	10	10 k	100 k	1.5 k	500	20	0.14
	3	100	10 k	1.0 M	1.5 k	100	3.0	0.7
	4	1000	1.0 k	1.0 M	0	10	3.0	5.2
5.7c	1	1	10 k	10 k	1.5 k	5.0 k	200	0.10
	2	10	10 k	100 k	1.5 k	500	20	0.14
	3	100	10 k	1.0 M	1.5 k	100	3.0	0.7
	4	1000	1.0 k	1.0 M	0	10	3.0	5.2
5.7d	1	A_{VOL}	0	∞	1.5 k	5.0 k	200	5.5
	2	A_{VOL}	0	∞	1.5 k	500	20	10.5
	3	A_{VOL}	0	∞	1.5 k	100	3.0	21.0
	4	A_{VOL}	0	∞	0	10	3.0	39.0
	5	A_{VOL}	0	∞	∞	0	3.0	—

particular frequency response, can be modified by the application of feedback. Knowing the internal details of the integrated circuit amplifier will enable the circuit designer to exploit it to maximum advantage as an operational amplifier.

As an example of this, Fig. 5.7a shows how one half of the dual

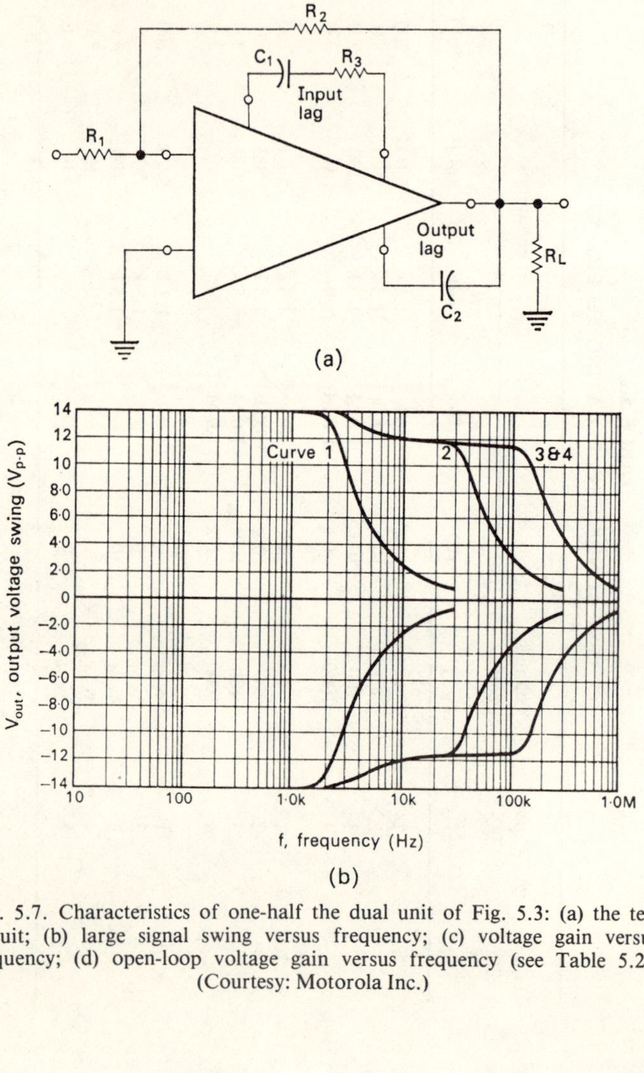

FIG. 5.7. Characteristics of one-half the dual unit of Fig. 5.3: (a) the test circuit; (b) large signal swing versus frequency; (c) voltage gain versus frequency; (d) open-loop voltage gain versus frequency (see Table 5.2). (Courtesy: Motorola Inc.)

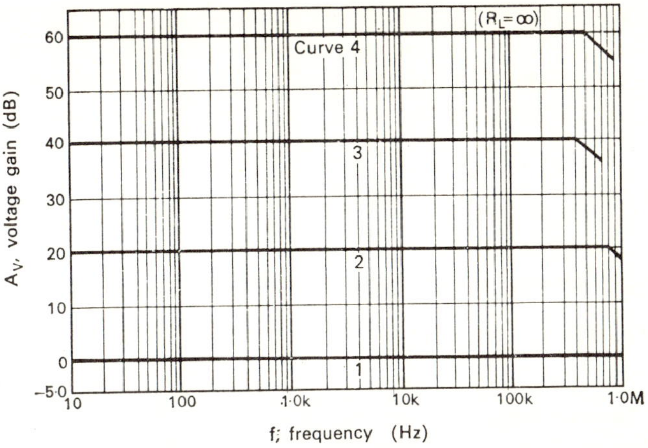

(c)

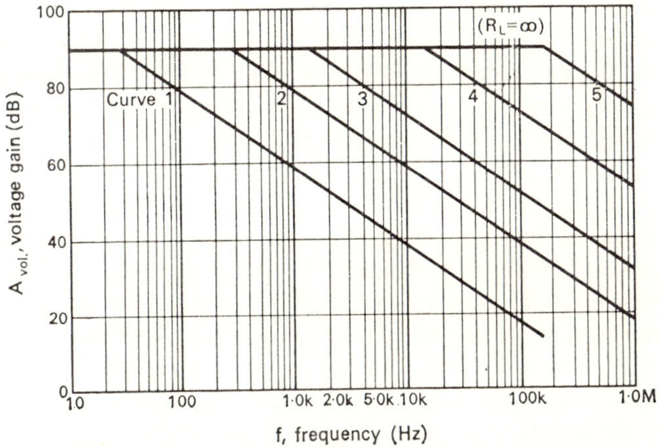

(d)

amplifier of Fig. 5.3 can be connected to obtain various output characteristics. Figure 5.7b shows how the *large-signal output voltage* versus *frequency* varies for various combinations of the components listed in Table 5.2. Figure 5.7c shows the *voltage gain* versus *frequency* curves, and Fig. 5.7d the *open-loop voltage gain* versus *frequency* curves.

CHAPTER 6

Metal–Insulator–Semiconductor Devices

6.1. Introduction

The metal-oxide–semiconductor transistor, or MOST, is one of the very important devices to be developed using solid-state technology. Although the basic idea is quite old (the first British patent for such a device having been awarded in 1935), it was only in 1964 that practical working devices were first produced. The main problems which had to be overcome were associated with the electronic properties of surfaces and interfaces, the interface being the region between the oxide and semiconductor.

The MOST is a field effect transistor (FET), and because there are other types of field effect transistor which do not rely on the use of an oxide insulator, the MOST is also referred to generally as an insulated gate field effect transistor or IGFET. As a final note on terminology, the abbreviation MIS device may be encountered, which stands for metal insulator semiconductor device. In early models an oxide was invariably used as an insulator, but more recently insulators such as silicon nitride have been used, so that the more general term MIS device is, in some ways, to be preferred.

Charge coupled Devices (CCDs) are a class of devices which depart quite markedly from conventional transistor operation. In these, field-effect control is applied through an insulated gate electrode as in the MOST, but the mobile charge is "clocked" along the semiconductor channel by means of a multiphase gate arrangement, as described in § 6.13.

6.2. The Basic MOST Mechanism

In Fig. 6.1a is shown a capacitor consisting of two parallel metal plates across which a voltage *V* is applied. For all practical purposes the resulting charge *Q* may be assumed to reside on the surfaces of the metal plates. In fact, the charge is distributed some way into the metal but because the free carrier (electron) density in a metal is high, of the

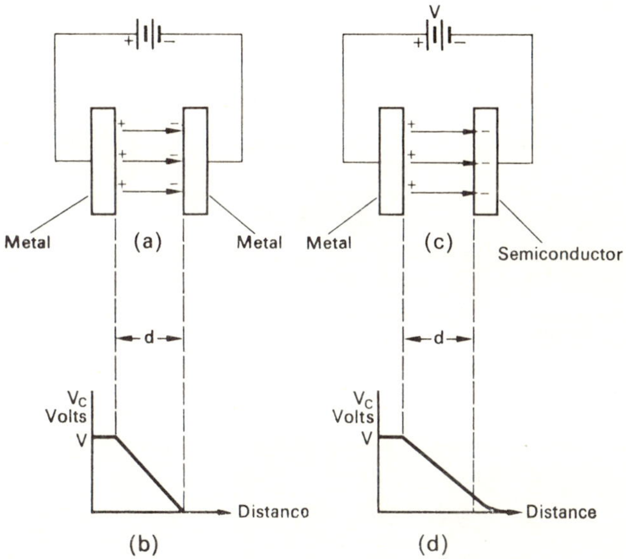

FIG. 6.1. The field-effect transistor: (a) a simple parallel-plate capacitor; (b) the voltage gradient graph for (a); (c) a parallel-plate capacitor with a semiconductor electrode; (d) the voltage gradient graph for (c).

order of 10^{22} per cubic centimetre, a very slight redistribution of the electrons within a few Angstroms of the true surface will accommodate the extra charge *Q* resulting from the applied voltage. (An Angstrom is 10^{-10} m.) The voltage as a function of distance can therefore be drawn as shown in Fig. 6.1b.

If now one of the metal plates is replaced by a semiconductor plate as shown in Fig. 6.1c then because the carrier density in the semiconductor is considerably less than that in the metal, of the order of 10^{16} per cubic centimetre, the electric field will penetrate the semiconductor to a considerably greater depth than in the metal. The voltage as a function of distance is then as sketched in Fig. 6.1d, where for simplicity the dielectric constant of the semiconductor is assumed to be equal to that of the dielectric between the plates. The voltage gradient within the semiconductor results in a redistribution of carrier density, and for the polarity of applied voltage shown in Fig. 6.1 the electron density increases towards the surface.

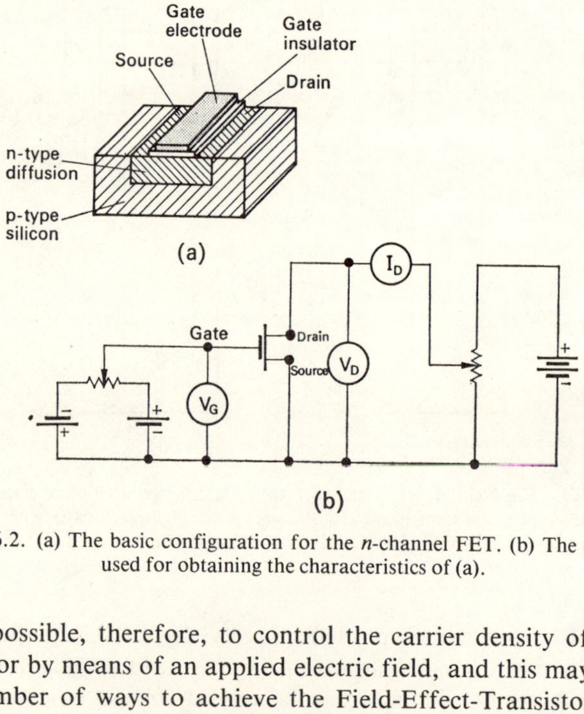

FIG. 6.2. (a) The basic configuration for the *n*-channel FET. (b) The circuit used for obtaining the characteristics of (a).

It is possible, therefore, to control the carrier density of a semiconductor by means of an applied electric field, and this may be used in a number of ways to achieve the Field-Effect-Transistor action. Figure 6.2a shows one method used in practice. The capacitor here consists of the metal plate termed the *gate* electrode, and the *n*-type semiconductor layer termed the conducting *channel*, and, of course, a dielectric between these two electrodes. Ohmic contacts are made to the channel at both ends, and these are termed the *source* and *drain* electrodes (one imagines current coming from the source electrode and leaving by the drain electrode). Figure 6.2b shows the symbol for the insulated-gate-field-effect-transistor (IGFET) and how bias voltages may be applied to obtain the device characteristics. Because the current carriers in the channel are of one type (and may be electrons or holes) the device is also known as a *unipolar* transistor.

Device Characteristics

The electric field within the conducting channel is determined by both the gate voltage, applied between gate and source, and the drain voltage, applied between drain and source. When the gate voltage is zero, i.e. gate and source at the same potential, there will still be mobile carriers throughout the channel because the semiconducting material is extrinsic *n*-type. Thus charge will flow from source to drain when the drain voltage V_D is applied. As V_D is increased from zero the current increases rapidly at first, but then levels off at comparatively high drain voltages. The reason for this is that at high drain voltages, the electric field between the drain *and gate* is high, and this depletes the region between gate and drain of carriers. Current is maintained because carriers are emitted from the undepleted region (source side of channel) into the depleted region, but this current is almost independent of drain voltage. The drain voltage-drain current characteristic with gate voltage zero is shown in Fig. 6.3a. The characteristic is divided into regions: (i) the *triode region*, so called because in the rigorous development of the theory the curve in this region can be predicted using a triode (i.e. three-electrode) model for the transistor; (ii) the *saturation region* where the current saturates as a function of drain voltage (and not to be confused with the saturation region discussed in connection with Fig. 4.4).

When the gate voltage is made positive the electron density in the channel is increased and therefore the current is increased over that obtained with zero gate voltage, for a given drain voltage. Apart from this, the curve follows the same general shape as shown in Fig. 6.3b.

With negative gate voltage the electron density is decreased, and the curves come below the zero gate voltage curve, Fig. 6.3b. As an aid to visualizing the action of the gate it may be noted that when the gate is positive extra electrons are attracted into the channel; when negative, some electrons are repelled from the channel.

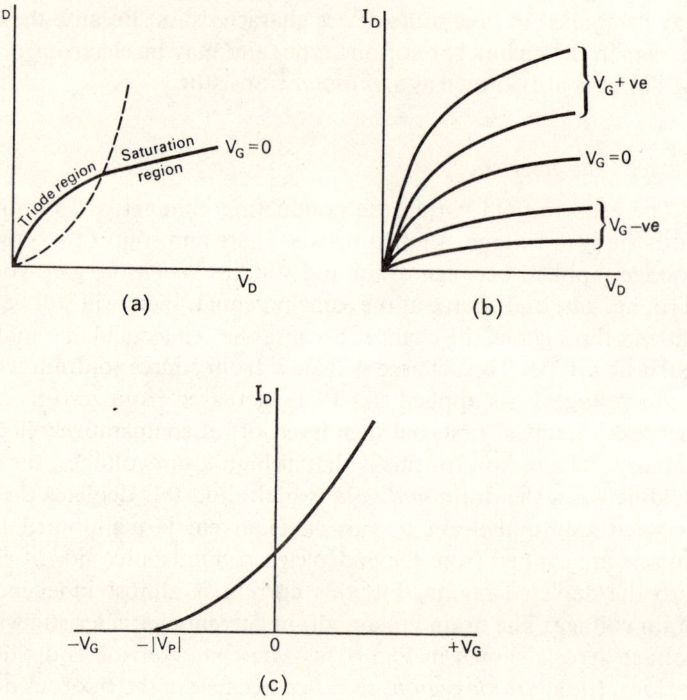

FIG. 6.3. Characteristic curves for the *n*-channel MOST of Fig. 6.2: (a) the output characteristic for $V_G = 0$, showing the triode and saturation regions; (b) the family of output characteristics; (c) the transfer characteristic in the saturation region.

A transfer characteristic for the device can also be plotted as shown in Fig. 6.3c. This is a graph of drain current against gate voltage, for a fixed drain voltage. A separate characteristic should be drawn for each value of drain voltage of interest, but it is seen from Fig. 6.3b that over the saturation range of drain voltage the current is reasonably constant (i.e. the saturation current) so that only the one curve is necessary to illustrate the shape of the characteristic. This strictly applies to the saturation region only, and separate curves should be drawn for drain voltages less than the first saturation value.

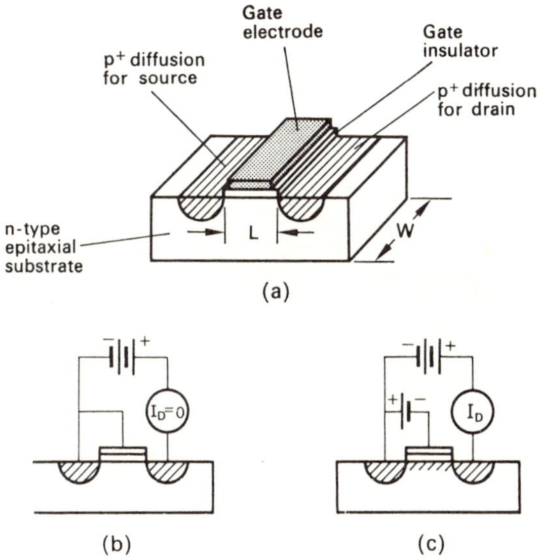

FIG. 6.4. (a) A *p*-channel enhancement mode MOST; (b) zero bias and no conducting channel between source and drain; (c) negative gate bias, forming a *p*-channel inversion layer between source and drain.

It will be seen from Fig. 6.3c that it is possible to make the gate voltage sufficiently negative to cut off the drain current; the value of gate voltage required to cut off is termed the *pinch-off voltage*, (V_P).

The type of transistor described is known as a *depletion-mode* transistor because it is possible to deplete the channel of carriers as well as increase (or enhance) the carrier density through an applied gate voltage. It is also possible to make transistors in which the carrier density can only be increased, or enhanced, by application of gate voltage and these are known as *enhancement-mode* transistors. Enhancement-mode transistors are especially desirable in logic circuits because ON/OFF switching is simply achieved by ON/OFF application of gate voltage. Enhancement-mode transistors are easier to make as *p*-channel devices (compared with *n*-channel) and only this type will be considered here.

Figure 6.4a shows a sketch of a *p*-channel enhancement-mode transistor. Two p^+ pockets are diffused into an *n*-type layer (which

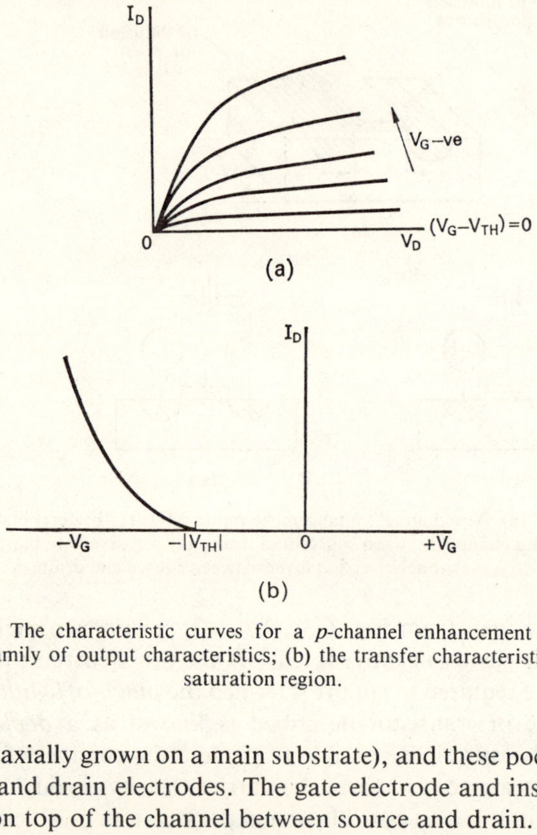

FIG. 6.5. The characteristic curves for a *p*-channel enhancement MOST: (a) the family of output characteristics; (b) the transfer characteristic in the saturation region.

may be epitaxially grown on a main substrate), and these pockets form the source and drain electrodes. The gate electrode and insulator are deposited on top of the channel between source and drain. When the gate voltage is zero, as shown in Fig. 6.4b, no conduction can take place between source and drain because effectively these form two *p–n* junctions connected back-to-back, one of which must be reverse-biased by the drain-source voltage. By making the gate negative with respect to the channel as shown in Fig. 6.4c, mobile negative carriers are repelled out of the channel at the same time that positive carriers (holes) are being drawn in. There comes a point when the positive carriers predominate and the surface of the *n*-type layer is said to be *inverted.* The inversion layer joins up with the two p^+ pockets as shown in Fig. 6.4c so that a *p*-type conducting channel is formed between source and drain. The energy band diagram is discussed in § 1.7.1.

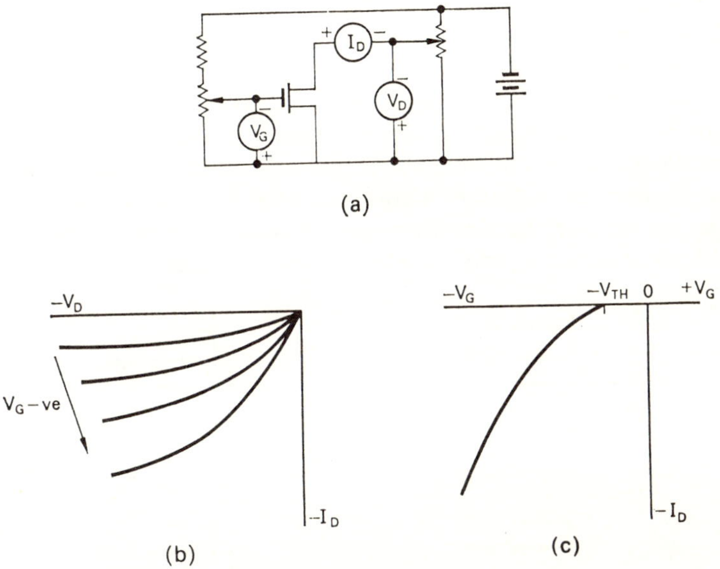

FIG. 6.6 A *p*-channel FET with negative drain voltage: (a) the bias connections using a single battery; (b) the output characteristics; (c) the saturation transfer characteristic.

In Fig. 6.5a is shown the output characteristics for a *p*-channel enhancement MOST, and in Fig. 6.5b, the transfer characteristic. These are drawn using the convention that positive current is the movement of positive charge into an electrode. Since the MOST transistor has a symmetrical structure it is only necessary to bias the drain positive with respect to source to obtain positive drain current I_D; this is true for both *p*- and *n*-channel devices. Also because of the symmetry of the device, the transistor is not limited to operating with positive drain voltage. Where a single battery supply is desired the drain may be made negative with respect to source as shown in Fig. 6.6a, which also enables a negative bias voltage to be applied to the gate. The characteristics then appear as shown in Fig. 6.6b and c.

6.3. MOST Symbols. Substrate Bias

N and P channel devices, which can be either enhancement or depletion mode, are available giving four basic types of MOST. Biasing of the substrate with respect to channel is also important, and the MOST is therefore a four-terminal device. The circuit symbols for the four basic types including substrate terminal are shown in Fig. 6.7. Recalling that the substrate will be extrinsically of opposite polarity to the channel the substrate arrow indicates the direction of forward bias current for the substrate-channel back junction (compare this with the standard symbol for a *p–n* junction diode).

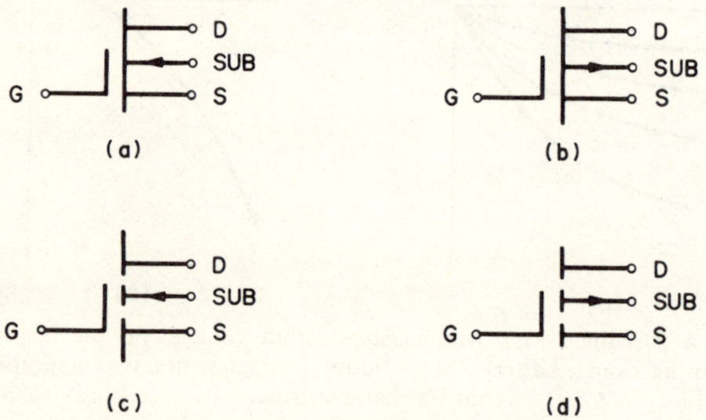

FIG. 6.7. MOST symbols: (a) *n*-channel depletion mode; (b) *p*-channel depletion mode; (c) *n*-channel enhancement mode; (d) *p*-channel enhancement mode.

The gate electrode symbol is usually (but not always) shown as an L on its side lying in line with the source terminal. For enhancement mode devices the channel is open circuit for zero bias, and this is indicated by means of the broken line between source and drain.

However, unless it is important to show these specific features the simpler MOST symbol introduced in Fig. 6.2b will be used.

Substrate bias has a marked effect on the transistor characteristics this being termed substrate degeneration. Where the substrate-channel

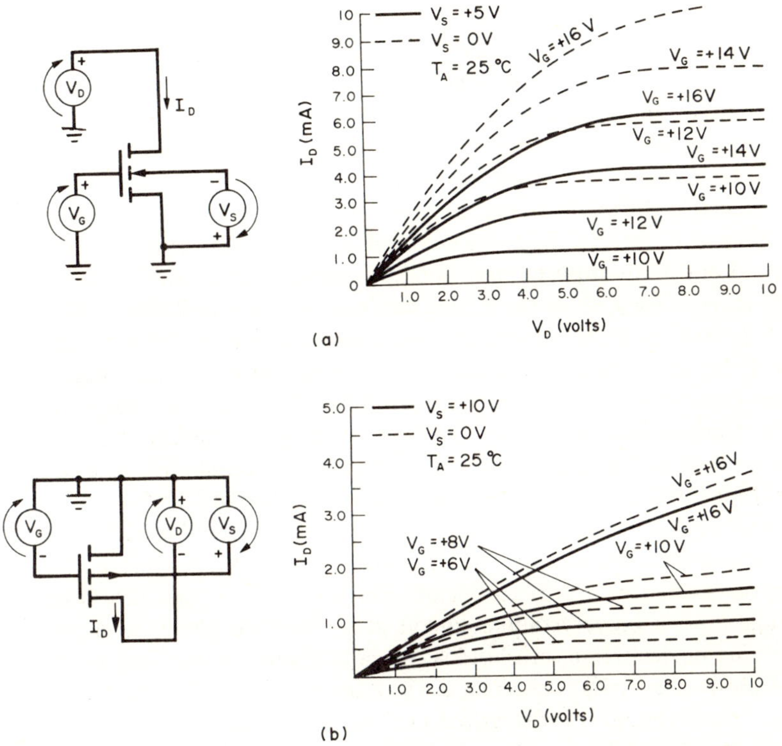

FIG. 6.8. Effect of substrate bias: (a) on an *n*-channel device, and (b) a *p*-channel device. (Courtesy: Motorola Semiconductor Products Inc.)

bias is such as to reverse bias the back junction as shown in Fig. 6.8, the depletion region of this junction produces an effective decrease in channel conductivity, which in turn results in a decrease in drain current as shown.

6.4. Threshold Voltage

It will be seen from the transfer characteristic that a threshold voltage (V_{TH}) must be applied to the gate before conduction can start. The threshold voltage for the enhancement-mode device corresponds to the pinch-off voltage for the depletion mode device. Threshold voltage depends on a number of factors the main ones being, the impurity doping (donor or acceptor density), the silicon surface ((111) or (100)), and the materials used for gate and gate insulator. Cleanliness of oxide is extremely important, great care being required in manufacture to prevent contamination especially from sodium ions which are particularly troublesome. In some cases, a phosphorous gettering layer is introduced in the oxide to neutralize active ions.

The metal–nitride–oxide–semiconductor (MNOS) transistor developed out of attempts to control oxide charge. Here, a thin ($\sim$20 Å) SiO_2 layer is formed on the silicon surface, and this is followed by a thick ($\sim$800 Å) Si_3N_4 overlayer. The gate electrode is deposited on top of the nitride. Unwanted charges in the insulator become trapped at the oxide-nitride interface thus stabilizing the unit. The charge remains trapped indefinitely but can be altered by application of a suitable level of voltage, and this has opened up the possibility of using MNOS transistors in erasable memories as described in § 12.5.

The threshold voltage is also a function of the material used for the gate electrode. Aluminium, which must be vacuum evaporated by electron-beam (see § 8.2) for cleanliness results in a high threshold voltage, and a polycrystalline silicon gate electrode results in a low threshold device. High threshold is typically about 3 V and low threshold about 1.5 V. In N-channel devices the gate "ON" voltage is of course positive and this drives the positive oxide charge to the oxide-silicon interface where it is most troublesome. As a result, N-channel devices are fabricated in the silicon (100) surface as the trapped charge density is reduced by a factor of about three to one compared to the more usual (111) surface.

6.5. The Polysilicon Gate

The polycrystalline silicon gate, as well as producing low threshold voltage, allows auto-registration of source and drain diffusions with respect to the gate. In this way, narrow channel transistors can be made in which the capacitive overlap between gate and source and drain electrodes is considerably less than in the aluminium gate transistor. The steps in the polysilicon gate process are sketched in Fig. 6.9. A thick oxide, approximately one micron, is grown to reduce parasitic coupling between top metallization and silicon substrate. Into this is etched a window, and a gate oxide, about 1000 Å thick is deposited. The polysilicon layer is then deposited over the complete surface and

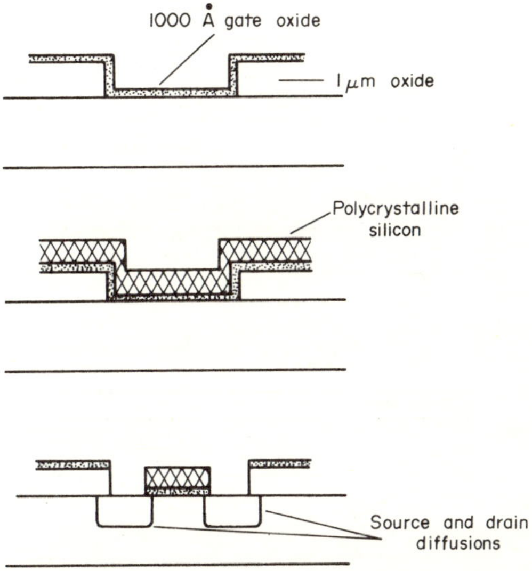

FIG. 6.9. Steps in fabricating the polysilicon gate.

etched to give the required conductor and gate pattern (polycrystalline material having a high carrier density). Following this step the gate oxide is etched to allow source and drain diffusions to take place (boron for *p*-channel devices, phosphorous for *n*-channel). It will be seen that the gate is self-aligning with the channel. The impurity will diffuse some way under the gate oxide, but the capacitive overlap created in this process is less by about 40 per cent than that of the aluminium gate device. Note that the impurity diffusion will also occur into the gate but this does not adversely affect device performance.

Polysilicon gates find special application in semiconductor memories (see Chapter 12). In one type of reprogrammable memory (§ 12.5) a polysilicon gate is formed within the oxide, as sketched in Fig. 6.10a. No external connection is made to the gate and hence it is referred to as a "floating-gate". Charge is transported to and from it through the thin underlying oxide from the silicon channel. Two methods are in

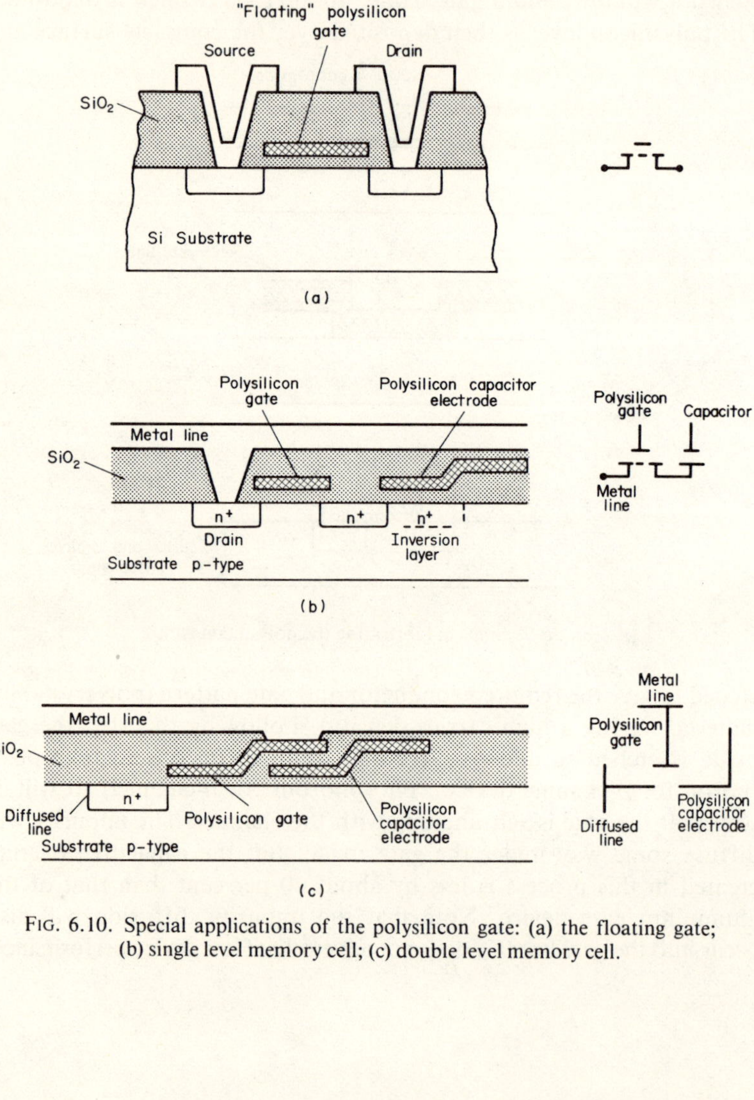

FIG. 6.10. Special applications of the polysilicon gate: (a) the floating gate; (b) single level memory cell; (c) double level memory cell.

use to generate the high energy electrons necessary to penetrate the oxide. In one, an avalanche field between channel and drain is used, and this also produces a fringing field which drifts the avalanche electrons into the gate. Here, the oxide beneath the gate may be about 1000 Å thick. In the other method, the oxide is very thin, in the range 20–35 Å, and a conventional metal gate is deposited on top of the thick oxide overlying the polysilicon gate. An electric field applied between the conventional gate and the substrate is sufficient to produce tunneling of electrons through the oxide, which then become trapped at the gate. In either case, the floating-gate remains negatively charged on removal of the charging mechanism, the charge remaining indefinitely. Thus the device forms a memory in a similar manner to the MNOS structure discussed in § 6.4.

Discharge of the floating-gate can be accomplished either by irradiation with ultra-violet light, or by application of an electric field, depending on the type of structure. Erasable memories utilizing these devices are described in § 12.5.

Polysilicon electrodes have also been used in the production of semiconductor memories of very high density (large scale integration or LSI). Fig. 6.10b and c illustrate the structures introduced by the Intel company. The memory cell in each case depends on the charge held by a polysilicon capacitor, which is charged (and discharged) through a field induced inversion channel. In the structure of Fig. 6.10b, the "switch" is a fairly conventional MOST but with a polysilicon gate buried in the oxide (note however, this is not a floating-gate, as it is externally connected in the normal way). Constructing the gate, and the capacitor upper electrode in this way allows space to be saved, as the metallization layer can be deposited on the top surface as shown. The capacitor in this cell has a capacitance of about 0.07 pF.

In the structure of Fig. 6.10c, space saving is taken a stage further by having the capacitor electrode come under the gate. This method of construction is referred to as a *double level polysilicon.* The capacitance in this case is about 0.03 pF. Here the gate-induced channel simply allows charge to move to and from the silicon under the capacitor electrode, and the action in fact is similar to that in the charge controlled devices described in § 6.7.

Memories utilizing these memory cells are described in § 12.4.

6.6. N-Channel Devices

During the initial development stages of MOS technology *p*-channel devices could be made more reliably, largely because the *p*-type diffusions into *n*-type substrates could more easily be controlled. However, as with most technological problems, there was sufficient incentive to solve the problems associated with fabricating *n*-channel devices, and *n*-channel integrated circuits as well as combined *n*- and *p*-channel devices, known as complementary MOS circuits, are now available. Complementary MOSTs are discussed in § 6.7.

n-channel logic circuits offer a number of advantages over *p*-channel circuits. For given dimensions they are faster because of the higher electron mobility. Surface mobility for electrons in silicon is in the range 400–470 cm^2/V-s, compared to 180–200 cm^2/V-s for holes. Also, parasitic conduction effects with *n*-channel devices are found to be less with the result that *n*-channel devices can be packed more closely together, this giving a higher density of devices on a given substrate size compared with *p*-channel devices. Typically, *n*-channel devices are made in a boron-doped (*p*-type) substrate. This is thermally oxidized to a thickness of about 1 μm, which is then etched to define gate and diffusion regions. Polysilicon gates and n^+ source and drain diffusions are then formed as described in the previous section.

6.7. Complementary Metal-Oxide-Semiconductor (CMOS) Devices

The basic complementary MOS unit consists of an *n*-channel MOST combined with a *p*-channel MOST in the same substrate. The constructional features are illustrated in Fig. 6.11. For the particular unit shown (which is made by the Motorola Company), the substrate is a lightly doped *n*-type (resistivity in the range 1–10 Ohm-cm), and the (100) surface is used. A 5000 Å oxide is thermally grown and etched to allow diffusion of the *p*-type "tubs" of about the same resistivity as the *n*-type substrate. The oxide is then removed and a new one grown, Fig. 6.11a. This is etched and p^+ diffusions made, Fig. 6.11b. Again the oxide is replaced with a new layer which is etched to allow the n^+ diffusions, Fig. 6.11c. Once again, a new oxide

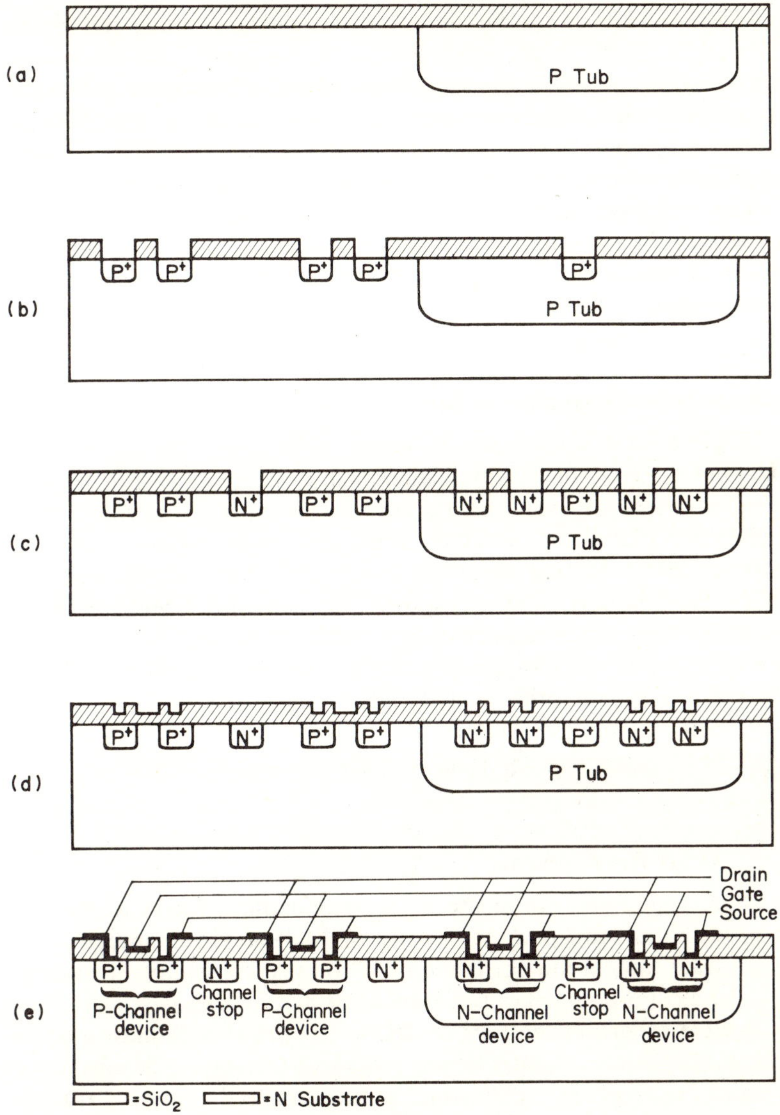

FIG. 6.11. Steps in fabrication of CMOS: (a) diffusing in *p*-type tubs; (b) diffusing in p^+ pockets; (c) diffusing in n^+ pockets; (d) gate oxide grown; (e) final CMOS structure.

is grown and etched for the source and drain contact areas, and for the gate areas. A gate oxide, about 1000 Å thick is grown in these exposed areas Fig. 6.11d, and subsequently etched out of the drain and source areas. Formation of the gate oxide is the most critical step. Metallization is then deposited over the complete surface and etched to the required conductor pattern, the final step being shown in Fig. 6.11e.

The "channel stop" diffusions are to prevent surface leakage between adjacent like-channel devices, since the surface between the drain of one and source of the other can become inverted under certain conditions.

6.8. Silicon on Sapphire (SOS)

Single crystal silicon can be grown as a thin film on sapphire (sapphire is a crystalline form of aluminium oxide Al_2O_3) which is a very good electrical insulator and a good heat conductor. The thin film is etched into islands into which the active devices can be fabricated in a manner similar to that described in § 11.5.2. for Gallium Arsenide FETs. SOS is particularly favoured for use with CMOS because parasitic capacitance between units is reduced to a negligible level, but it is also used to advantage with other technologies such as *n*-channel. As well as sapphire, other insulating substrates are being tried, one of these being Spinel ($MgAl_2O_4$), also a gem-stone, which is easier to machine than sapphire and is a better crystallographic match to silicon.

6.9. Other Developments

Other techniques are in use by various manufacturers to try and improve speed and packing density of MOS devices especially for large scale integration (LSI) required for semiconductor memories, and the most prominent of these will be described briefly.

Double Diffusion MOS or D–MOS. The principle is illustrated in Fig. 6.12a. Here, a lightly doped *p*-type surface layer, denoted as a π region, receives an n^+ diffusion for the drain, and a *p*-type diffusion which is used to define the channel and which receives the n^+ source diffusion. The effective channel length is as shown in Fig. 6.12a, and is the length between the n^+ source and the π region connecting with the drain. As shown by eqn. (6.1) in § 6.10, the gain-bandwidth product is inversely proportional to the square of the channel length. With D–MOS technique, channel lengths of the order of 1 μm can be achieved compared with 5 μm for conventional silicon gate technology.

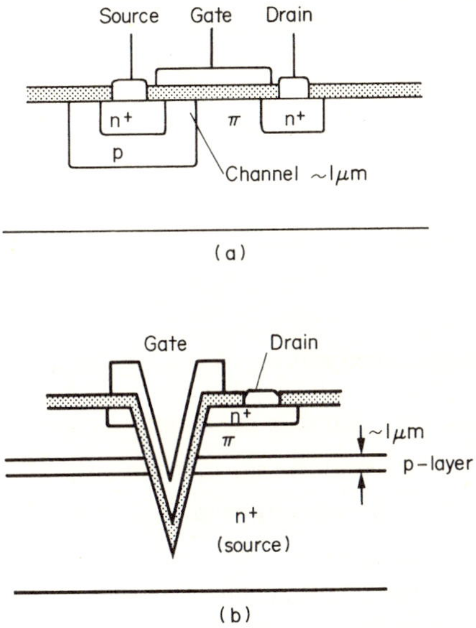

FIG. 6.12. (a) Double diffusion (D-MOS) to obtain a narrow channel; (b) V-notch (V-MOS) to obtain a narrow channel.

V–MOS Fig 6.12b. The electrical arrangement is similar to D–MOS but the narrow *p*-type channel is obtained structurally in quite a different manner. An n^+ substrate forms the source (which is therefore common to all devices made on the same substrate), and a *p*-type layer of about 1 μm forms the channel. A π region makes contact between the *p*-type channel and the n^+ drain. As shown in Fig. 6.12b, the field-effect exercised by the gate takes place along the sides of a *V*-notch etched into the structure.

6.10. Equations for the MOST

The voltage–current relationships in the MOST can be described quite accurately by means of equations derived under certain simplifying assumptions. Using the same assumptions it is also possible to relate the mutual conductance g_m to the physical parameters of the device. The derivation of these equations is too involved to be reproduced here, and only the results will be stated which are useful in describing the behaviour of the device. It is necessary to divide the operating characteristics of the MOST into two distinct regions, the *triode region*, and the *saturation region*, (see Fig. 6.3a), and the dividing line is formally defined by the condition that for saturation to exist, $|V_D|>|V_G-V_{TH}|$ for an enhancement-mode device, and $|V_D|>|V_G-V_P|$ for a depletion-mode device. By writing V_0 for V_P or V_{TH}, the condition can be stated generally as:

$$\text{saturation conditon: } |V_D|>|V_G-V_0|\cdot$$

Mutual conductance is defined in the usual way as:

$$g_m=\frac{\text{Small change in drain current}}{\text{Small change in gate voltage}}$$

with drain voltage held constant.

The equations relating mutual conductance to the physical parameters of the device can be shown to be:

$$\text{triode region:} \qquad g_m=\mu_{FE}\frac{C_g}{L^2}|V_D|.$$

$$\text{Saturation region:} \quad g_m=\mu_{FE}\frac{C_g}{L^2}|V_G-V_0|, \tag{6.1}$$

where C_g = gate-to-channel total capacitance (F); L = channel length; (m); μ_{FE} = field effect mobility, (m^2/V sec); V_D = drain-source voltage, (V); V_G = applied gate-source voltage, (V); V_0 = threshold voltage (V_{TH}) for enhancement mode devices and pinch-off voltage (V_P) for depletion mode devices, (V); g_m = mutual conductance, Siemens.

For example, typical values might be:

$$C_g = 10\ \text{pF};\ \mu_{FE} = 200\ \text{cm}^2/\text{V sec};\ L = 10\ \mu\text{m},$$

so that for $V_D = 1$ V, the g_m in the triode region is:

$$g_m = \frac{200 \times 10^{-4} \times 10 \times 10^{-2}}{10^{10} \times 10^{-12}}\ \text{Siemens.}$$

$$= \underline{\underline{2.0\ \text{millisiemens.}}}$$

Equation (6.1) shows that g_m is proportional to C_g, but increasing C_g will adversely affect the high-frequency performance of the device. A more useful criterion for judging performance is the gain × bandwidth product. This is the product of the maximum voltage gain ($G = g_m R_L$, where R_L is the load resistance) and the 3 dB bandwidth $B = \dfrac{1}{2\pi C_g R_L}$, and is given by:

$$G \times B = \text{gain} \times \text{bandwidth product} = \frac{g_m}{2\pi C_g}. \qquad (6.2)$$

Substituting for g_m from the eqn. (6.1) gives:

$$\text{triode region:} \qquad G \times B = \frac{\mu_{FE}}{2\pi L^2} \, |V_D|,$$

$$(6.3)$$

$$\text{saturation region:}\ G \times B = \frac{\mu_{FE}}{2\pi L^2} \, |V_G - V_0| .$$

The current–voltage equations can also be shown to be:

$$\left.\begin{aligned} &\text{triode region:} && |I_D| = \frac{C_g \mu_{FE}}{L^2}\left(|V_G - V_0|\,|V_D| - \tfrac{1}{2}|V_D|^2\right),\\ &\text{saturation region:} && |I_D| = \frac{C_g \mu_{FE}}{2 . L^2}\,|V_G - V_0|^2 . \end{aligned}\right\} \quad (6.4)$$

One result of the simplified analysis is that eqn. (6.4) for the saturation region shows the drain current to be independent of drain voltage, i.e. the a.c. slope resistance of the device in saturation to be infinite, whereas in practice it is finite and ranges from 20 kΩ to 200 kΩ.

Equation (6.4) shows a quadratic relationship exists between drain voltage and drain current in the triode region, and, as already mentioned, the current is independent of drain voltage in the saturation region (but only approximately so). A quadratic relationship also exists between drain current and gate voltage when the device is operating in saturation.

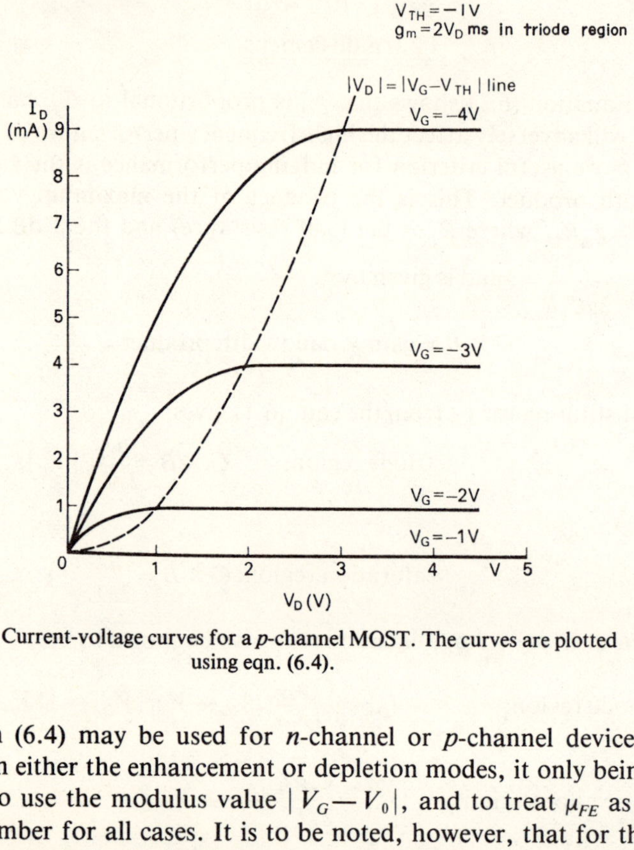

FIG. 6.13. Current-voltage curves for a *p*-channel MOST. The curves are plotted using eqn. (6.4).

Equation (6.4) may be used for *n*-channel or *p*-channel devices, operating in either the enhancement or depletion modes, it only being necessary to use the modulus value $|V_G - V_0|$, and to treat μ_{FE} as a positive number for all cases. It is to be noted, however, that for the depletion-mode device $|I_D|$ is finite for $V_G = 0$, while for the enhancement-mode device, $|I_D|$ is zero for $V_G = 0$, and these restrictions must be taken into account when using eqn. (6.4). The curves for a *p*-channel enhancement-mode device are plotted in Fig. 6.13.

6.11. The MOST as a Resistor

The channel resistance of a MOST in saturation is very high compared with the diffused resistors used in bipolar integrated circuits; typically a value of the order of 20 kΩ per square can be achieved compared with 200 Ω per square for the diffused resistor, and the higher value is achieved using considerably less area than the diffused resistor. In Fig. 6.14a a MOST is shown connected in a manner to ensure saturation for all drain voltages. This is so, since saturation occurs

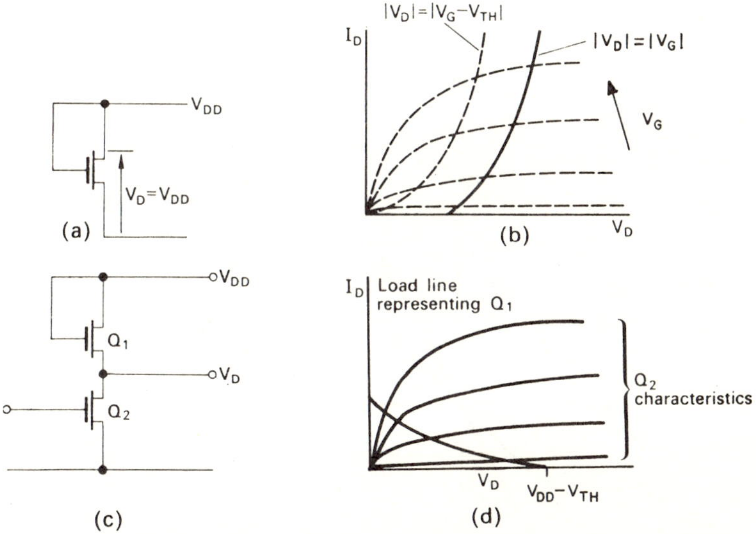

FIG. 6.14. The MOST in saturation as a load resistor: (a) the biasing connection to ensure saturation; (b) the operating line for the MOST connected as (a); (c) the MOST load used with a MOST driver stage; (d) the characteristic curves and load-line for the circuit of (d).

for all drain voltages V_D greater than $V_G - V_{TH}$, and the method of connection automatically makes $V_D = V_G$. The locus of the operating point for varying V_D is shown in Fig. 6.14b and for comparison the line for $V_D = V_G - V_{TH}$ is also shown.

When used as a load resistor the circuit is as shown in Fig. 6.14c; the corresponding output characteristics and load-line are shown in Fig. 6.14d. This circuit is discussed more fully in Chapter 7.

More recently, switching circuits similar to that of Fig. 6.14c have been introduced in which the load is a depletion mode transistor. In this case the gate is connected directly to the source of the load so that the device operates along the zero gate-source voltage characteristic. In the high speed Gallium Arsenide FET inverter described in § 11.5.2, both driver and load transistors operate in depletion mode.

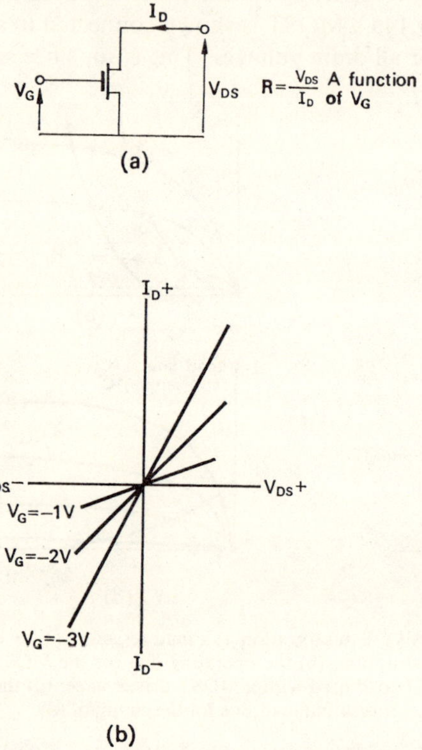

FIG. 6.15. The MOST as a voltage-controlled resistor: (a) the resistive circuit; (b) the voltage-controlled resistor curves.

The MOST may also be used as a voltage controlled variable resistor by confining the operation to the triode region. Figure 6.15 shows the circuit connection, and Fig. 6.15b shows the resistor characteristics. It will be seen that by operating near the origin, the resistance, which is the slope of the V_{DS}/I_D line, can be altered by using V_G to shift from one line to another, (and, of course, V_G can be continuously variable so that operation is not limited to the discrete lines shown in Fig. 6.15b). It will also be observed that because of the symmetry of the MOST structure, the characteristics are symmetrical about the origin; in effect, source and drain connections are interchangeable.

6.12. Input Impedance of a MOST

The input impedance is simply that presented by the gate–semiconductor capacitance with the gate insulator as dielectric. The d.c. resistance in general tends to be very high, in the range 10^{12}–10^{15} Ω. Use is made of the high input time-constant in dynamic memory systems (see § 12.6). The a.c. impedance should be the d.c. resistance in parallel with the input capacitance, (and for this reason, increasing C_g to increase the g_m results in a lower input impedance). In practice it is also found that the series resistance of the gate lead can have a marked effect causing the effective Q-factor (and hence the equivalent a.c. impedance) to decrease rapidly with increase in frequency. Q-factors as low as 20 may be encountered unless special precautions are taken to minimize the effects of series lead resistance, as would be required, for example, in MOSTs intended for v.h.f. operation.

6.13. Charge Coupled Devices (CCDs)

In its simplest form the CCD is rather like an elongated, multiple gate, IGFET, as shown in Fig. 6.16. The field effect control exercised by the various gates enables "charge packets" to be moved around within the silicon channel, these packets being induced into the channel from a source via an input signal control gate. Discrete charge packets, representing the input signal level at different times, can exist simultaneously at different locations in the CCD and this is what gives the CCD its signal processing capabilities. Both analog and digital signals can be processed, and some applications are described in § 7.10 and 12.6. The gate clocking waveforms ϕ_1, ϕ_2, ϕ_3 necessary to ensure charge movement in one direction will be described following the description of the general features illustrated in Fig. 6.16.

Two main types of CCD are available, the surface-channel CCD (SCCD), and the buried-channel CCD (BCCD). The SCCD, Fig. 6.16a, operates in the enhancement mode, that is, the charge packets consist of minority carriers which form an inversion layer where they are located. The charge packets are transferred along the Si–SiO_2 interface and the advantage of this is that the capacitance formed by the charge

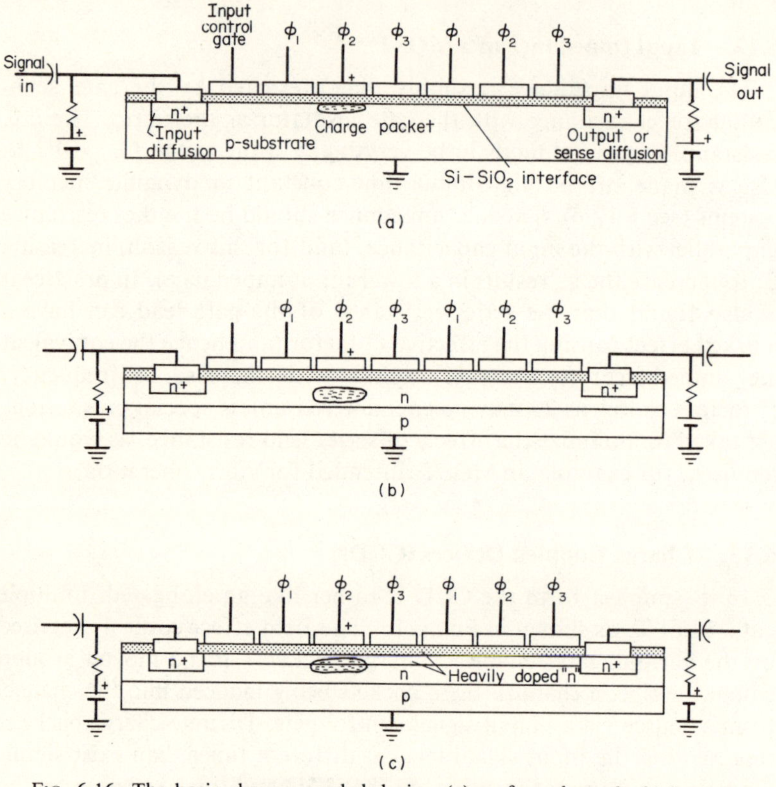

FIG. 6.16. The basic charge coupled device: (a) surface channel; (b) buried channel; (c) peristaltic (twin-channel) mode.

packet and the particular gate is high since only the oxide separates them, and hence the charge handling capacity is high. One major disadvantage of the surface mode is that surface trapping states reduce both the charge transfer efficiency and the speed since they can behave as recombination centres for electrons and holes, and, where they do release electrons after trapping, the time constant can be long.

In the BCCD, the transfer channel is shifted to within the bulk of the silicon and this eliminates the effects of surface trapping. In the model shown in Fig. 6.16b, an *n*-type layer is grown on the *p*-type substrate, and it is reverse biased through the output (or sense) diffusion. The bias potential on the *n*-channel is greater than that on the gates so that a depletion layer forms along the Si–SiO_2 interface as well as along the substrate *p–n* junction. These two depletion layers confine the charge packets to a channel which is about 3 μm beneath the Si–SiO_2 interface, and thus they avoid the surface trapping states. The disadvantage of the buried channel mode is that the capacitance between the charge packet and any particular gate is reduced because the spacing is increased, and hence the charge handling capacity is reduced (in practice by about a factor of 10). Note that majority carriers are involved in the BCCD.

It is also important to look at the mechanism of charge transfer for both SCCDs and BCCDs. Common to both is a self-induced electric field which originates from the charge packet and which results in the initial large drift of electrons into the "potential well" under the nearest positive gate. This field will decrease as the packet gets smaller until finally, thermal diffusion takes over for the small amount of remaining charge at the end of a given transfer. Thermal diffusion is a relatively slow process and imposes a limitation on speed (in practice it limits the maximum gate frequencies to about 10 MHz; surface trapping states will reduce this further). Externally induced drift fields also exist in the channel, which arise from the potential difference between adjacent gates, but they are only significant in the bulk. This is because equipotential surfaces exist at the interface immediately under the gates whereas within the channel a potential gradient exists, acting in the direction of transfer. The externally induced drift field dominates over thermal diffusion in the bulk so that total charge transfer is speeded up.

The *profiled peristaltic CCD* (or PCCD), combines the advantages of both modes of operation without their attendant disadvantages, this device having been invented by L. J. M. Esser of the Philips Research Laboratories, Eindhoven. The PCCD is similar to the BCCD of Fig. 6.16b, with the important modification that the *n*-type layer is very highly doped *n*-type in a thin layer at the surface. The term "profiled" is used to describe this variation in doping. The thin *n*-type surface layer prevents the charge packet interacting with surface states, but at the same time, with appropriate gate bias it is possible to enhance

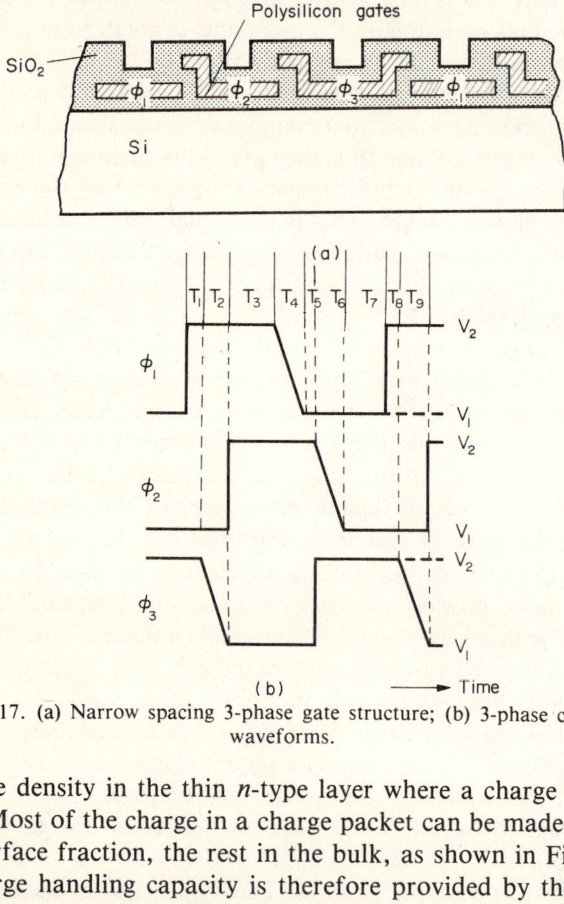

FIG. 6.17. (a) Narrow spacing 3-phase gate structure; (b) 3-phase clocking waveforms.

the charge density in the thin *n*-type layer where a charge packet is present. Most of the charge in a charge packet can be made to reside in this surface fraction, the rest in the bulk, as shown in Fig. 6.16c. High charge handling capacity is therefore provided by the surface mode, while fast removal of the last of the charge is provided by the bulk mode (this dominating the transfer time); the combined operation is referred to as the twin channel mode. The term *peristaltic* is used to describe the way in which the charge packets dilate and contract in the transfer process which is similar to the way in which food is transferred through the digestive system.

In experimental results quoted by Esser, gate clocking frequencies greater than 135 MHz have been used, and transfer efficiencies of 99.999 per cent have been achieved (transfer efficiency gives the percentage of the charge actually transferred from one gate to another in a single transfer operation).

There are many refinements in structural detail omitted from Fig. 6.16 for clarity. For example, in order to reduce spacing between gates (this should not be greater than 2–3 μm), advantage is taken of the self-alignment possible with polysilicon gates (§ 6.5). Fig. 6.17a illustrates a three-phase gate structure in which the separation between gates is the oxide thickness (unfortunately the method requires a large number of processing steps). Figure 6.17b shows the waveshapes for a three-phase clocking (ϕ_1, ϕ_2, ϕ_3 of Fig. 6.16) which results in maximum charge handling capacity. Applying this to the SCCD of Fig. 6.16(a), V_2 would be well above, and V_1 just slightly above, the surface inversion threshold voltage. Assuming a charge packet present, the sequence of events would be: period T_1, charge is shared under adjacent ϕ_1, ϕ_3 gates; period T_2, the charge under ϕ_3 gate is transferred to under the adjacent ϕ_1 gate; period T_3, charge is shared under adjacent ϕ_1, ϕ_2 gates; period T_4, charge under ϕ_1, gate is transferred to under adjacent ϕ_2 gate; period T_5, charge held under ϕ_2 gate; period T_6, charge under ϕ_2 gate is transferred to under adjacent ϕ_3 gate; period T_7, charge held under ϕ_3 gate; period T_8 charge is shared under adjacent ϕ_3, ϕ_1 gates; period T_9 charge under ϕ_3 gate is transferred to under adjacent ϕ_1 gate. Clocking systems other than three-phase are also in use. Input and output arrangements also vary in detail depending on application. A floating gate arrangement (§ 6.5) has been used at the output as a very sensitive detector of charge.

Finally, it should be noted that signal information cannot be stored indefinitely in a CCD. Thermally generated minority carriers will tend to accumulate under the "empty" gates in the surface channel mode, while leakage and charge transfer inefficiency will reduce the charge in charge packets. A method of regenerating, or refreshing, binary data is described in § 7.10.

CHAPTER 7

MOS Circuits

7.1. Introduction

The main advantages of the metal oxide semiconductor transistor for use in integrated circuits are that it requires considerably less area than the bipolar transistor, and, it can be used as a high value load resistor thus making possible integrated circuits using MOSTs only. These circuits can be highly complex; for example some commercially available circuit functions contain 500 MOSTs on a single silicon chip measuring 70 by 70 mil (1 mil = 0.001 in.) which represents a component packing density of 10^5 per square inch.

The disadvantages of the MOST compared with the bipolar type are mainly associated with lower speed. The field effect mobility is generally low compared with the drift mobility which makes the MOST slow compared with the bipolar transistor. Also, with present fabrication methods, the minimum channel length for the MOST is limited to about 5 μm (1 $\mu m = 10^{-6}$ m), and some overlap between gate electrode and source and drain electrodes is unavoidable which results in high stray input capacitance. Both these limitations (comparatively long channel length, and high stray input capacitance), again limit speed, and therefore also limit the high-frequency performance.

Improvements in fabrication methods are being made which should improve the high frequency (or speed) performance of the MOST.

N-channel devices are faster than *P*-channel because surface mobility for electrons ranges from 400–470 cm^2/V-s, compared to 180–200 cm^2/V-s for holes. Power dissipation is also a critical consideration and a figure of merit often used is the Power-Delay Product (usually

TABLE 7.1.
MAJOR LSI TECHNOLOGIES (1976)

Technology	Propagation delay (ns)	Power-delay product (pJ)	Density		Chip size (mm^2)
			(Devices/mm^2)	(Gates/mm^2)	
High-threshold *p*-channel metal gate	80	450	150	50	7 × 7
p-channel silicon gate	30	145	270	90	6.5 × 6.5
n-channel silicon gate	15	45	285	95	6 × 6
n-channel silicon gate depletion-load	12	38	320	107	6 × 6
n-channel double-polysilicon	10	35	525	175	6 × 6
Silicon-gate C-MOS	10	0.5	220	45	5.5 × 5.5
V-MOS D-MOS	5	20	600	225	—
SOS/C-MOS	2–5	0.1	650	275	5 × 5
I^2L (double level)	5–50	0.01–1	500	150	5.5

(Courtesy: *Electronics* and L. Altman.)

given in picojoules). Table 7.1 lists some values, and a detailed comparison will be found in the article by Laurence Altman, pages 74–81, *Electronics*, April 1, 1976. A basic inverter circuit utilizing Gallium Arsenide FETs operating in the depletion mode is described in § 11.5.2.

7.2. The Basic Inverter (or Negate) Circuit

In Fig. 7.1 is shown the basic MOST inverter circuit in which transistor Q_1 is used as a load resistor (see § 6.11). The behavior of the circuit can be determined using the current eqns. (6.4). The load transistor always operates in saturation therefore the saturation current equation applies. The driver transistor Q_2 operates through saturation into its triode region therefore the two regions must be considered.

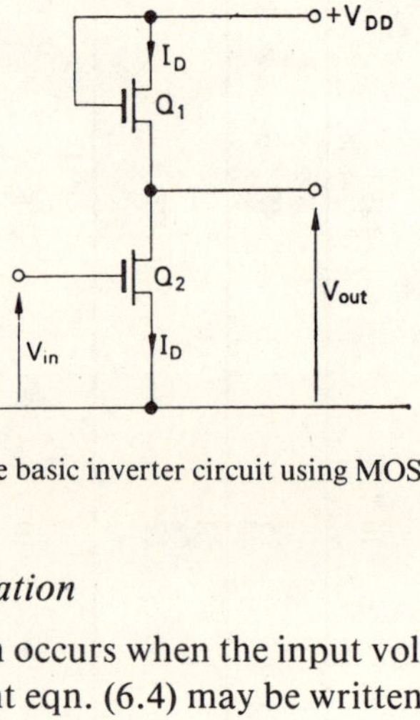

FIG. 7.1. The basic inverter circuit using MOSTs.

Driver and Load in Saturation

This mode of operation occurs when the input voltage V_{in} is low, i.e. $|V_{in}| \ll |V_{DD}|$. The current eqn. (6.4) may be written as:

$$I_D = \frac{\beta}{2}(V_G - V_0)^2, \tag{7.1}$$

where

$$\beta = \frac{C_g \cdot \mu_{FE}}{L^2} \tag{7.2}$$

and the other symbols are defined in connection with eqn. (6.4). Note that β is a constant for any particular device. In particular, for the load device, $V_G = (V_{DD} - V_{out})$, and, denoting the parameters of the load transistor by the subscript 1, the current equation for the load becomes:

$$I_D = \frac{\beta_1}{2}(V_{DD} - V_{out} - V_{01})^2. \tag{7.3}$$

Denoting the driver transistor parameters by the subscript 2, the current equation for saturation is:

$$I_D = \frac{\beta_2}{2}(V_{in} - V_{02})^2. \tag{7.4}$$

Since the two transistors are in series, eqns. (7.3) and (7.4) can be set equal to each other:

$$\frac{\beta_1}{2}(V_{DD} - V_{out} - V_{01})^2 = \frac{\beta_2}{2}(V_{in} - V_{02})^2,$$

from which,

$$V_{out} = (V_{DD} - V_{01}) - \sqrt{\left(\frac{\beta_2}{\beta_1}\right)} \cdot (V_{in} - V_{02}). \tag{7.5}$$

Equation (7.5) relates the output voltage V_{out} to the input voltage V_{in} when both devices are in saturation, and is used to plot the first half of the voltage transfer characteristic (Fig. 7.2).

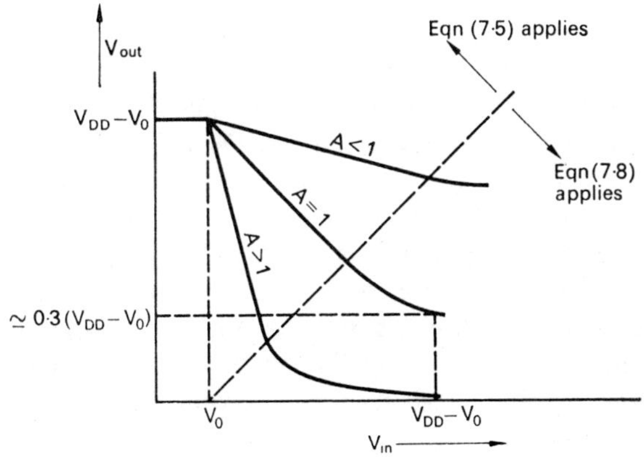

FIG. 7.2. The voltage transfer characteristic for the circuit of Fig. 7.1.

The voltage gain of the stage is the ratio of change in output voltage to change in input voltage, which from eqn. (7.5) is:

$$\text{voltage gain} = A = \frac{\text{change in output voltage}}{\text{change in input voltage}}$$

or

$$A = -\sqrt{\left(\frac{\beta_2}{\beta_1}\right)}. \qquad (7.6)$$

The negative sign shows that the output voltage change is in anti-phase to the input voltage change. Also, it is seen that to obtain large gains the driver β should be very much greater than the load β.

Driver in Triode Region, Load in Saturation

Equation (7.3) still applies to the load device but the triode current equation must be used for the driver thus:

$$I_D = \beta_2 \left[(V_{in} - V_{02}) \,.\, V_{out} - \tfrac{1}{2} V_{out}^2\right]. \qquad (7.7)$$

Equations (7.3) and (7.7) may be set equal to each other to give:

$$\frac{\beta_1}{2}(V_{DD} - V_{out} - V_{01})^2 = \frac{\beta_2}{2}\left[(V_{in} - V_{02}) \,.\, V_{out} - \tfrac{1}{2} V_{out}^2\right]$$

or

$$(V_{DD} - V_{out} - V_{01})^2 = 2A^2\left[(V_{in} - V_{02}) \,.\, V_{out} - \tfrac{1}{2} V_{out}^2\right] \qquad (7.8)$$

This is a quadratic equation which relates V_{out} to V_{in}, and is plotted to complete the voltage transfer characteristics (Fig. 7.2). The detailed manipulation of the equation is straightforward but tedious, and only some results will be quoted here. When the voltage gain A is close to zero the output voltage is nearly equal to V_{in} which in turn is $(V_{DD} - V_0)$. Since this is also the value of the output voltage in the OFF state, the switching action is poor. When A is very large, i.e. $A_2 \gg 1$, the output voltage approaches zero when the input voltage is applied and so the switching action is good. For intermediate values of A, the transfer characteristic will be intermediate between poor and good. The transfer characteristics for three values of A are sketched in Fig. 7.2.

When an inverter gate is built in integrated circuit form the substrate is common to both load and driver devices as shown in Fig. 7.3a.

With V_{in} off, V_{out} rises towards $(V_{DD}—V_{TH})$. This reverse biases the load channel with respect to substrate, and substrate degeneration occurs as discussed in connection with Fig. 6.8. In effect, the load channel is constricted as a result of the reverse bias and the load gate voltage V_{GL}, Fig. 7.3(a), does not have to drop to the threshold level V_{TH} in order to cut-off the load channel. The output voltage V_{out}, does not rise as high as $(V_{DD}—V_{TH})$, Fig. 7.3(b).

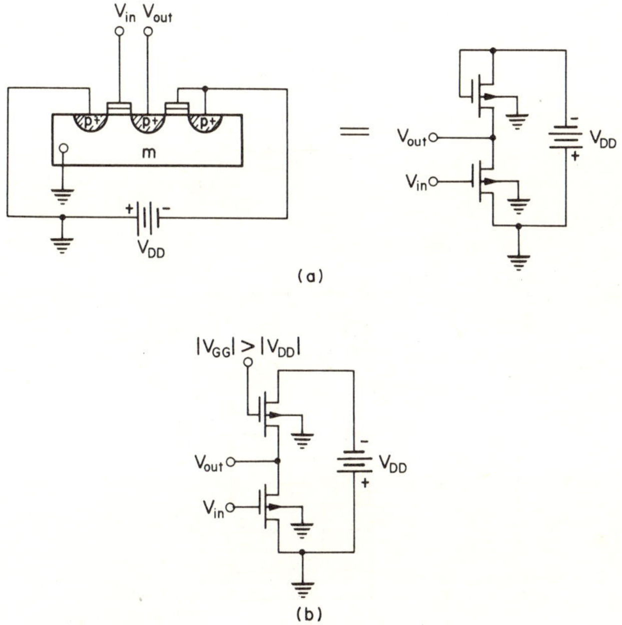

FIG. 7.3. (a) Integrated inverter layout. (b) Operating characteristics showing reduction in V_{out}. (c) Use of separate gate supply for load.

The difficulty can be overcome by returning the load-gate to a separate bias supply as shown in Fig. 7.3c and by ensuring that the separate supply $|V_{GG}|$ is sufficiently greater than $|V_{DD}|$ to compensate for the modulation effect. However, in the following circuits the load-gate will be shown returned to V_{DD} as the principle of operation of the circuit is the same.

7.3. The Basic NOR Gate

A circuit for a three-input NOR gate using MOSTs is shown in Fig. 7.4a, and a possible layout in integrated circuit form in Fig. 7.4b.

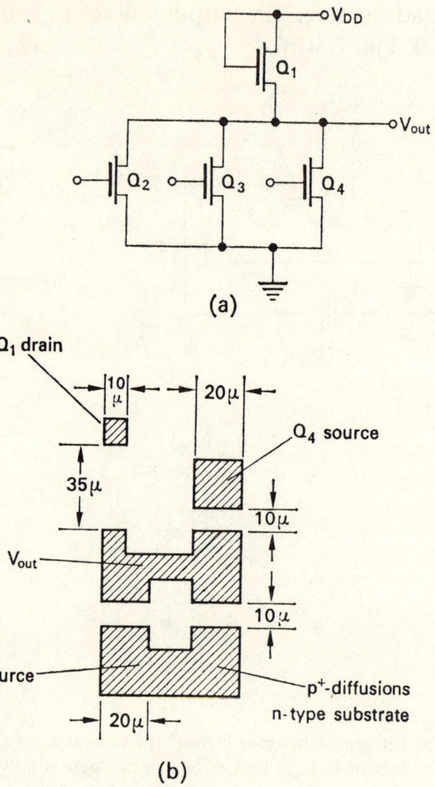

FIG. 7.4. The basic MOST NOR gate: (a) the circuit for a three-input gate; (b) a possible integrated circuit layout.

For clarity only the p^+-diffusion pockets in the n-type substrate are shown. To complete the physical circuit insulating layers are deposited over the channels, followed by the gate electrodes. These are designed to cover the complete channel length with minimum overlap at source and drain electrodes. Metallic connections are deposited at the same time as the gate electrodes but these make direct contact with required p^+-regions.

The NOR gate is seen to be three inverter stages in parallel and sharing a common load transistor Q_1. When all the driver transistors, Q_2, Q_3, and Q_4 have zero gate input, i.e. at logic level $\underline{0}$, the output is at logic level $\underline{1}$. When logic level $\underline{1}$ is applied to any one of the driver transistor gates the transistor conducts and the output drops to logic level $\underline{0}$. Hence the NOR function (see § 4.2) is realized. For an n-type substrate with p^+-diffusions for source and drain electrodes, negative logic (see § 4.1) must be used, i.e. V_{DD} must be negative with respect to ground.

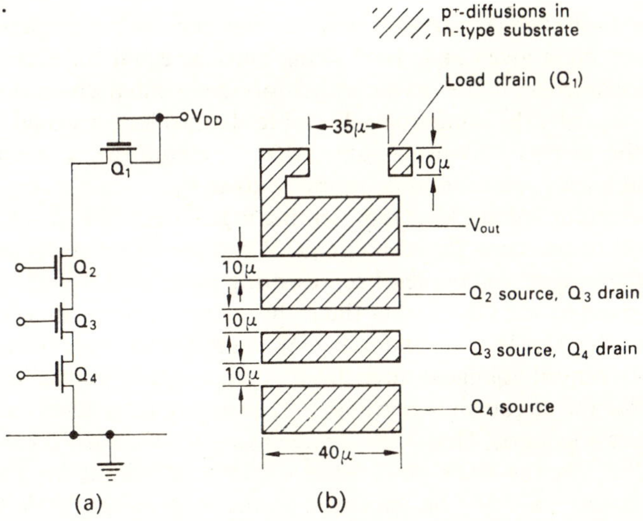

FIG. 7.5. The basic MOST NAND gate: (a) the circuit for a three-input gate; (b) a possible integrated circuit layout.

7.4. The Basic NAND Gate

A circuit for a three-input NAND gate using MOSTs is shown in Fig. 7.5a along with a possible integrated circuit layout, Fig. 7.5b. Again, for clarity, only the p^+-diffusions in the n-type substrate are shown. The driver transistors Q_2, Q_3, and Q_4 are connected in series with the load transistor Q_1. When all the driver transistors have their gates at logic level $\underline{1}$ (again, using negative logic) the series chain of transistors is conducting and the output voltage is therefore at logic level $\underline{0}$. This is the NAND function. When any one of the driver gates is at logic level $\underline{0}$ the transistor is driven into cut-off which effectively open-circuits the series chain and the output level rises to V_{DD} or logic level $\underline{1}$.

7.5. Comparison of Areas Required for the Basic NAND and NOR Gates

The basis for comparison will be that the output voltage swing representing a given logic level swing must be equal for each circuit. It is assumed that the driver transistors have equal channel lengths since normally the minimum achievable channel length would be used for either circuit. It is also assumed that the circuits work at equal ON current levels, and have equal supply voltage V_{DD}.

The output voltage level representing logic level $\underline{1}$ is $V_{DD} - V_{01}$, and this will be the same for each circuit. When the NOR gate is switched ON, the output voltage level representing logic level $\underline{0}$ is the voltage drop given by the ON current times the source-drain resistance of the driver transistor that is conducting. When the NAND gate is switched ON, the output voltage is three times as great as that for the NOR gate since for the circuit shown in Fig. 7.5, there are three identical driver transistors in series. Hence the voltage *swing* for the NAND gate is less than that for the NOR gate. Since the channel widths are fixed the NAND gate can only be improved to the level of the NOR gate by increasing the widths of the driver transistors such that the total ON resistance of the series driver chain is reduced so as to be equal to that of any one of the driver transistors of the NOR gate. The NAND circuit therefore requires greater area than the NOR circuit and for this reason the NOR circuit is to be preferred.

For either type of circuit, the difference between the W/L ratios for load and driver transistors should be noted. The eqn. (7.2) for β can be written as:

$$\left.\begin{aligned}\beta &= \frac{C_g \cdot \mu_{FE}}{L^2},\\ &= \frac{C \cdot LW \cdot \mu_{FE}}{L^2},\\ &= C \cdot \frac{W}{L} \cdot \mu_{FE},\end{aligned}\right\} \tag{7.9}$$

where C = gate–channel capacitance per unit area, W = channel width, L = channel length.

Substituting eqn. (7.9) in eqn. (7.6) for voltage gain gives:

$$A = -\sqrt{\left(\frac{W_2/L_2}{W_1/L_1}\right)}. \tag{7.10}$$

Thus, for high voltage gain, the ratio W_2/L_2 for the driver transistor should be considerably greater than W_1/L_1, for the load transistor. For the circuit shown in Fig. 7.4b $W_1/L_1 = 2/7$, $W_2/L_2 = 2$, and therefore the voltage gain (modulus) is 2.65.

7.6. NAND–NOR Gate

Combined gates are readily achieved using MOSTs and a basic circuit is shown in Fig. 7.6a. A possible integrated circuit layout is shown in Fig. 7.6b, where again, for clarity, only the p^+-diffusions in the n-type substrate are shown. The layout shows clearly the greater area required for a single NAND driver compared with the NOR driver transistor.

From the circuit of Fig. 7.6a it can be seen that either the complete NAND chain, *or* the transistor Q_5 must be ON to switch the output to logic level $\underline{0}$. Since the circuit performs the Negate function (inverter) automatically, it is termed a NAND/NOR gate.

7.7. Small-signal Analog Amplifier

The advantages of the MOST as a small-signal amplifier compared with the bipolar transistor, are that it has a very much higher input resistance, of the order of $10^{10}\,\Omega$ compared with $10^4\,\Omega$, and it produces less distortion. Its main disadvantage is the limitation on its high-frequency performance for the reasons discussed in § 7.1, and it is expected that these will disappear with improvements in fabrication techniques and the introduction of new materials. It is interesting to note that some of the trapping effects which cause a reduction in the field-effect mobility are less effective at high frequencies due to the finite time-constants associated with them.

In order to bring out the basic points of amplifier operation the low-frequency behaviour will be considered. The parameters which are most useful for this type of analysis are (i) the mutual conductance g_m, (ii) the drain-source slope resistance or a.c. resistance r_{ds}, and (iii) the

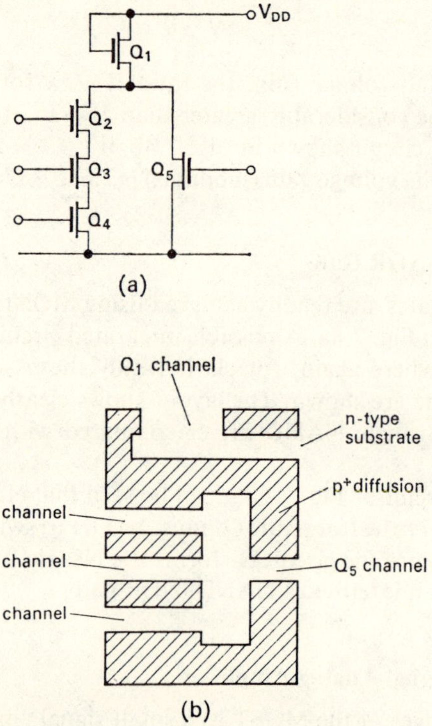

FIG. 7.6. The basic MOST NAND–NOR gate: (a) the circuit for a four-input gate; (b) a possible integrated circuit layout.

voltage amplification factor μ. These parameters may be defined in terms of small changes in the device voltages and current:

$$g_m = \frac{\text{small change in drain current}}{\text{small change in gate voltage}} \tag{7.11}$$

with drain voltage held constant.

$$r_{ds} = \frac{\text{small change in drain voltage}}{\text{small change in drain current}} \tag{7.12}$$

with gate voltage held constant.

$$\mu = -\frac{\text{small change in drain voltage}}{\text{small change in gate voltage}} \tag{7.13}$$

for drain current held constant.

The small changes in voltages and currents can be signal variations (which for most cases of interest will be continuous variables, e.g. sinewaves) and these may be denoted by i_d for a small change in drain current; v_d for a small change in drain voltage; and v_g for a small change in gate voltage.

The relationship between the parameters is found as follows. From eqn. (7.11) the variation in drain current due to variation in gate voltage alone is $i_d' = g_m v_g$, and from eqn. (7.12) the variation in drain current due to variation in drain voltage alone is $i_d'' = v_d/r_{ds}$. Therefore the *total* change in drain current due to both variations in drain voltage and gate voltage is:

$$\begin{aligned} i_d &= i_d' + i_d'', \\ &= g_m v_g + v_d/r_{ds}. \end{aligned}$$

By arranging for this total change to be zero, which is the condition for which μ is defined, then:

$$i_d = 0 = g_m v_g + v_d/r_{ds}$$

$$-v_d/v_g = g_m \cdot r_{ds}$$

or

$$\mu = g_m \cdot r_{ds} \tag{7.14}$$

on substituting eqn. (7.13) for the left-hand side, $-v_d/v_g$. Equation (7.14) is very useful and important in amplifier theory.

The basic circuit for the common source amplifier is shown in Fig. 7.7a. The name common-source is used because the source is common to both input and output. From the circuit of Fig. 7.7a it is seen that:

$$v_d = -v_L$$
$$= -i_d \cdot R_L,$$

and since $i_d = g_m v_g + v_d/r_{ds}$, this can be substituted to give:

$$v_d = -R_L(g_m v_g + v_d/r_{ds}).$$

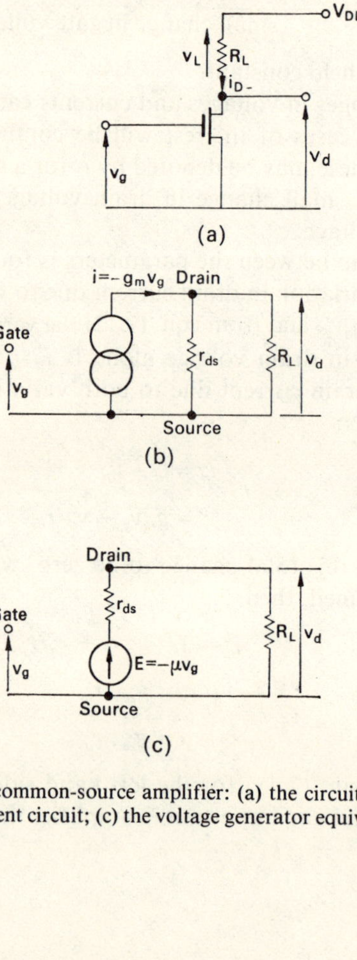

FIG. 7.7. The basic common-source amplifier: (a) the circuit; (b) the current generator equivalent circuit; (c) the voltage generator equivalent circuit.

Collecting terms and rearranging:

$$-g_m v_g = v_d(1/R_L + 1/r_{ds}). \tag{7.15}$$

Equation (7.15) can be interpreted in terms of a current generator supplying a current of $-g_m v_g$ to two resistors r_{ds} and R_L in parallel, and across which a voltage v_d is developed. Since R_L is external to the transistor, it can be further deduced that the transistor itself is represented by the current generator and r_{ds}. The equivalent circuit is shown in Fig. 7.7b. The voltage gain of the stage as determined from the equivalent circuit is:

$$A = \frac{v_d}{v_g} = -g_m R_L' \tag{7.16}$$

where R_L' is the parallel combination of r_{ds} and R_L.

Equation (7.16) can also be rearranged, using the relationship in eqn. (7.14) to give:

$$A = -\frac{\mu \,.\, R_L}{r_{ds} + R_L}. \tag{7.17}$$

This leads to the equivalent voltage generator circuit shown in Fig. 7.7c. Note that in both equivalent circuits, v_d and not v_L appears across the load resistor R_L.

The transistor is normally operated in the saturation region for small-signal amplification, therefore r_{ds} will be high. If this is much greater than the load resistance the expression for voltage gain, eqn. (7.16), approximates to $A \simeq -g_m R_L$.

The g_m in the saturation region is given by eqn. (6.1), where V_{TH} can be replaced by the more general term V_0. It will be seen that there is a dependence of mutual conductance on the gate voltage but this can be kept small in practice if so desired.

The dependence of g_m on gate voltage is a useful feature of the MOST as it enables the gain to be readily controlled through an applied voltage. By altering the fixed bias V_G the g_m is varied directly as shown by eqn. (6.1) and hence the gain is varied directly as shown by eqn. (7.16).

7.8. Large-signal Analog Amplifier

The voltage gain for the circuit of Fig. 7.1 is given by eqn. (7.6). Although this is derived for switching applications no restriction is placed on the voltage changes which may take place; they may therefore be analog signals. The only restriction is that both transistors operate in saturation, a condition which is easily met in practice. The voltage gain is given by eqn. (7.10) and the remarkable feature which this brings out is that the voltage gain is dependent only on the channel dimensions of the two transistors. One very important consequence of this is that the gain can be made independent of temperature over a wide range of temperature.

7.9. MOSTs as Frequency Doublers

The basic circuit for a frequency doubler is shown in Fig. 7.8 and such a circuit is easily constructed in integrated circuit form since the

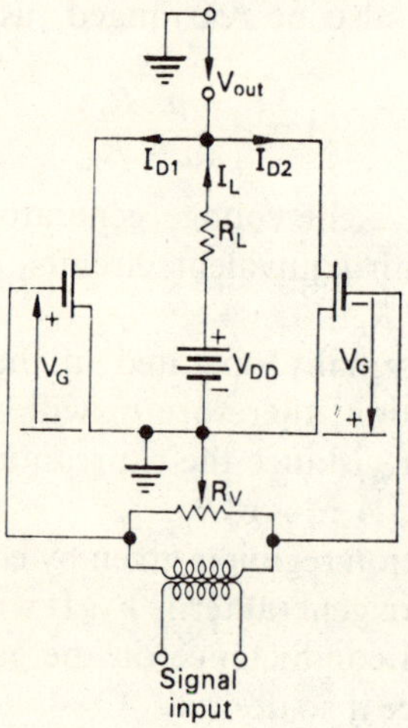

FIG. 7.8. A frequency-doubler circuit using MOSTs.

two sources are common and the two drains are common. For correct operation, the circuit must be balanced, that is, the input voltage must be divided equally to the two gates, and this can be achieved by adjusting the variable resistor R_v; also the transistors must be matched, which means that their respective values of β and V_0 must be equal. Assuming this can be achieved, the total load current is given by:

$$I_L = I_{D1} + I_{D2}$$

Applying the saturation equation (6.4) to the individual drain currents, but noting that since the gates are fed in anti-phase, the gate voltage on one is V_G at the same time that it is $-V_G$ on the other:

$$I_L = \frac{\beta}{2}(V_G - V_0)^2 + \frac{\beta}{2}(-V_G - V_0)^2,$$

$$= \beta\,(V_G^2 + V_0^2). \qquad (7.18)$$

If the gate voltage V_G can be represented by a sine-wave, e.g. $V_{\max}\sin(2\pi ft)$, where f is the frequency of the wave, then:

$$I_L = \beta\,[V_{\max}^2 \sin^2(2\pi ft) + V_0^2]$$

$$= \beta\,\{V_{\max}^2[\tfrac{1}{2} - \tfrac{1}{2}\cos(2\pi ft)] + V_0^2\}. \qquad (7.19)$$

Equation (7.19) shows that the total load current contains a signal component at twice the frequency of the input wave.

7.10. CCD Applications

Digital applications include shift registers, discussed in § 12.6 and logic functions such as NAND and NOR gates. Signal regeneration (to combat the signal degeneration discussed in § 6.13) is easily introduced in digital systems. One arrangement is illustrated in Fig. 7.9a. Charge packets arriving at the sense diffusion of the first CCD are arranged to bias the second CCD input to below threshold so that the digital signal in the second CCD is the inverse of the first signal. In the absence of charge packets from the first CCD, the input gate to the second CCD is biased just above threshold and charge packets are induced in it from its input diffusion. The signal inversion is of no consequence (this is a Negate, or NOT operation) since it is easily reinverted if required. What is important is that the charge packets in the second CCD are independent of the amplitude of the charge packets arriving from the first CCD, providing of course these have not been allowed to degrade to the point where there is ambiguity between presence and absence of a packet. Digital information can be recirculated indefinitely in this manner.

NAND and NOR functions can be incorporated in the CCD structure, and a possible arrangement for a NAND gate, along with

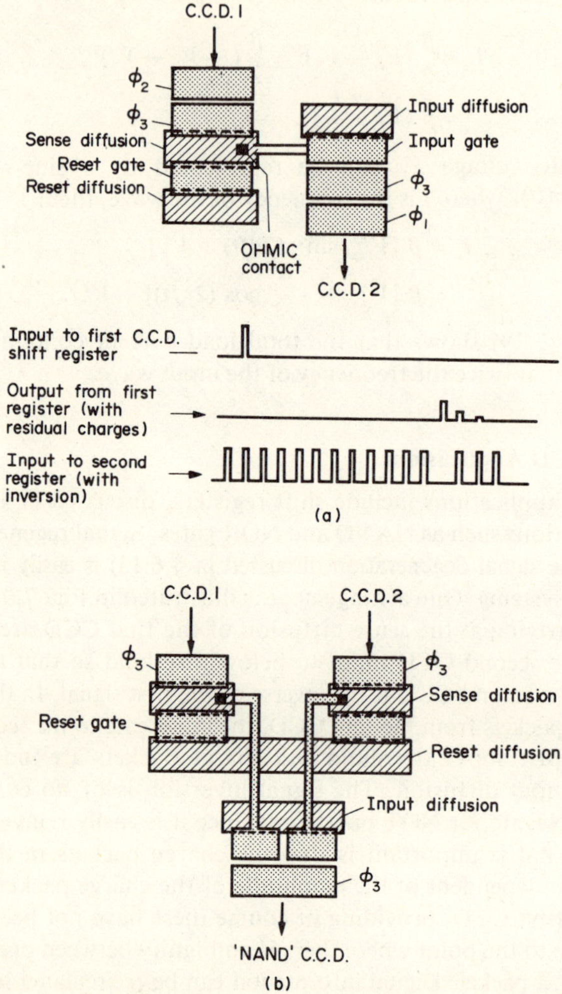

FIG. 7.9. CCD applications: (a) signal refreshing; (b) NAND gate. (Courtesy: *The Radio & Electronic Engineer* and J. D. E. Beynon.)

regeneration is shown in Fig. 7.9b. Charge packets arriving simultaneously at the sense diffusions of CCD1 and CCD2 together bias the input gate to the third CCD to below threshold, so that charge regeneration coupled with inversion occurs as described for the arrangement of Fig. 7.9a. The input gate to the third CCD is seen to be in half-sections, and one or other of these alone, biased below threshold, is not sufficient to prevent charge transfer from the input diffusion.

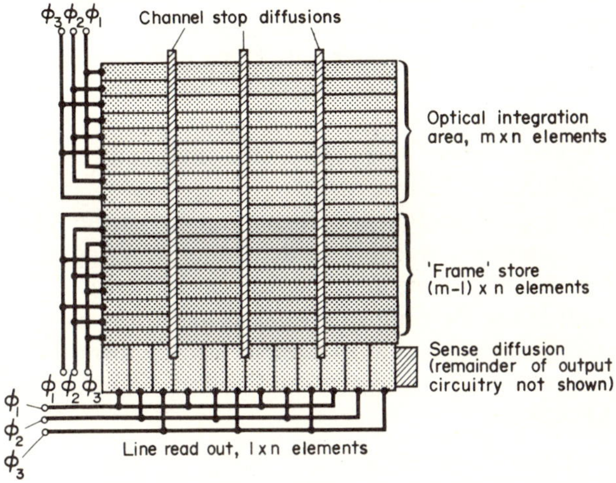

FIG. 7.10. CCD imager. (Courtesy: *The Radio & Electronic Engineer* and J. D. E. Beynon.)

CCDs have been used in various analogue applications. For example, they may be used to introduce a time delay. The analogue signal is sampled at the Nyquist sampling rate and the samples fed into a CCD. With careful circuit design the size of the charge packets can be made proportional to the sampled voltage levels. Time delays may then be introduced which are a function of the clocking frequency, using the shift register principle; delays in the range 10^{-5}s to 10^{-1}s have been achieved.

The fact that light energy can be used to generate hole-electron pairs in silicon opens up possibilities in optical imaging. Figure 7.10 illustrates one possible arrangement. Polysilicon gates, which are transparent to light are used in the Optical Integration Area. Light penetrates through the gates and the gate insulator layer to generate hole-electron pairs just within the silicon surface, the density of the hole-electron pairs being proportional to the light intensity. Elements in the same row in the Optical Integration Area are isolated from each other by means of channel stop diffusions (similar to those described in § 6.7). Clocking signals to the imaging area are periodically "frozen" so that optically generated minority carriers can collect under the gates. For example, phase ϕ_1 may be held constant at a value above threshold while ϕ_2 and ϕ_3 are below threshold; minority carriers will collect under the ϕ_1 gates, the density under any particular gate being proportional to the light intensity in that region. When clocking is recommenced the charge packets under the ϕ_1 gates are shifted down the array and into the frame store and read-out line. The frame store and read-out line are optically shielded.

Clocking signals transfer the information out of the read-out line. When this is emptied, clocking to the frame store moves the frame information down one line and line read-out is repeated. During the period required for the complete frame read-out the clocking signals to the optical imaging area are frozen as already mentioned to allow minority carriers to be generated and collected (i.e. an integration process takes place).

7.11. The CMOS Inverter

The basic advantage of CMOS is that power consumption can be kept low because in the steady state (either ON or OFF) the CMOS inverter draws only leakage current. The constructional details of a CMOS inverter are shown in Fig. 7.11(a) (fabrication steps are described in section 6.7), and the equivalent circuit in Fig. 7.11(b). Capacitance C represents the output capacitance, and is important in determining power consumption (as well as speed). When a positive logic $\underline{0}$ is applied to the gate input, Q_1, being *n*-channel turns off, and Q_2, being *p*-channel, turns on. The steady state equivalent circuit is shown in Fig. 7.11(c). Q_2 is represented by a comparatively low resistance, of the order of 1–2 kΩ, while Q_1 is virtually an open circuit, the resistance being of the order of 1000 MΩ. The output capacitance charges through Q_2 and during this part of the switching transient power is dissipated in Q_2. The output voltage is seen to rise to V_{DD} which is logic level $\underline{1}$.

If now, the input logic is switched from 0 to positive logic 1, Q_2 goes open circuit, and Q_1 a low resistance, so C discharges through Q_1. In doing so power is again dissipated, this time in Q_1. The new steady state equivalent circuit is shown in Fig. 7.11(d), where the output voltage is seen to drop to 0. It is only during the switching transient that significant power is dissipated as already described; in either steady state condition, the inverter is seen to be a virtual open-circuit, with only leakage current contributing to the power dissipation.

As with other Metal Oxide Semiconductor Transistors, the input gates of the CMOS circuit are protected with Zener diodes to prevent voltage breakdown, the gate oxide layers being very easily damaged by voltage spikes. These diodes are fabricated as part of the integrated circuit and require little additional space.

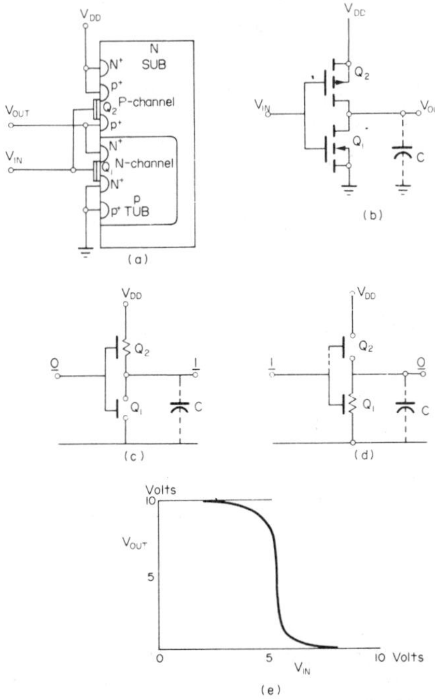

FIG. 7.11 (a) A CMOS inverter layout. (b) The equivalent circuit. (c) and (d). The equivalent circuits for the steady state switched conditions. (e) The transfer characteristic.

CHAPTER 8

Thin-film Circuits

8.1. Introduction

Thin films are widely used in microelectronic circuitry to provide (a) interconnecting conductor patterns, (b) highly accurate and stable resistors, and (c) capacitors. Thin-film transistors can also be made but these are not available commercially to the same extent that junction transistors and MOSTs are. The factors behind the comparatively slow development of the thin-film transistor would appear to be economical rather than technological. In production, the MOST can be made utilizing much of the equipment and techniques already available for junction transistor production. With thin-film transistors extensive additions to plant would be required and new techniques would have to be mastered on a production basis. None the less, the thin-film transistor (or TFT) has a number of unique advantages. It is directly compatible with thin-film components already widely in use; also it is extremely flexible in design as a variety of semiconducting materials may be used for the conducting channel.

Thin films range from about 100 Å to about 1 μm in thickness. Thick films, described in Chapter 9, range from about 10 μm upwards, and the range 1–10 μm may be classed as either thick or thin. There is no sharp dividing line between thick and thin films and in practice the distinction is usually based on the method of manufacture.

8.2. Thin-film Deposition Methods

The most widely used methods for thin-film deposition are (i) vacuum evaporation, and (ii) sputtering. Using either of these methods, films may be deposited on insulating substrates, usually glass or ceramic.

Vacuum Evaporation

A basic system for vacuum evaporation is illustrated in Fig. 8.1. The substrate which eventually will carry the thin film circuit is placed in a bell-jar along with a source of the material which has to be deposited as a thin film. The bell-jar is evacuated down to a pressure of about 10^{-5} torr or less, (1 torr = 1 mm of mercury), and the source material heated so that it vaporizes. The vapour condenses on the inside surfaces of the bell-jar, including the substrate, to form a thin film.

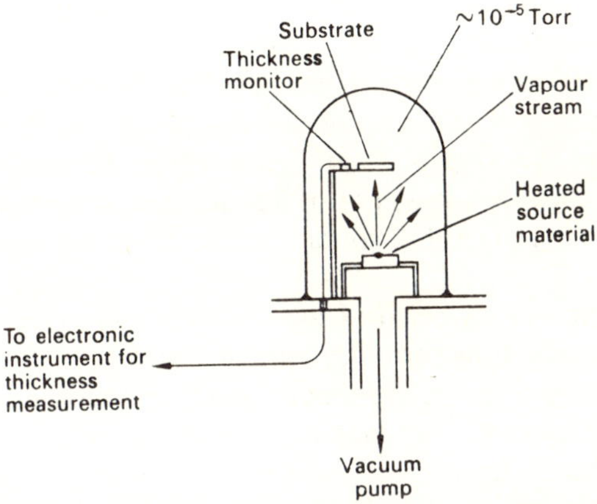

FIG. 8.1. A basic thin-film vacuum evaporation system.

The bell-jar has to be evacuated of air and other gases otherwise the vapour stream would collide with gas molecules. This would both reduce the rate of film deposition on the substrate, and gas molecules would be carried onto the substrate to act as impurities in the final film. The vapour stream is emitted as a spherical cloud and therefore to obtain a reasonably uniform thickness of film the substrate should not be too close to the source. The separation between source and substrate is typically in the range 18–25 cm. This is also sufficient to prevent the source heating the substrate.

The evaporant material may be heated directly by placing it on a resistance heater. The resistance heater can be made of stranded tungsten wire wound as a spiral and on which is hung loops of the evaporant material in wire form. Electric current is passed through the tungsten filament, heating it, and causing the loops to evaporate. Where the evaporant is in powder or lump form, the heater is usually made of tantalum or molybdenum in the shape of a shallow crucible or "boat" in which the evaporant is placed. Indirect heating may also be employed in which a quartz crucible is used to carry the evaporant and around which the resistance heating wire is wound. Resistance heating is only suitable for evaporating low melting-point materials, e.g. melting point temperatures less than about 900°C, and it suffers from the serious disadvantage that the heater invariably gives off impurities which are deposited with the required film. For high melting-point materials such as valve metals and oxides, the most suitable form of heating is by means of electron-beam bombardment. With this method, temperatures in excess of 2500°C can readily be achieved. Also, the heat can be localized to a spot of about 3 mm diameter on the actual evaporant material so that contamination from the source is virtually eliminated.

Where compounds, such as indium antimonide (InSb) have to be obtained in thin film form a particular problem occurs in that the elements making up the compound tend to separate when the compound is heated, and they may not reform in the correct proportions as a film (when in the correct proportions the compound is known as a stoichiometric compound). High-power electron-beam bombardment can often overcome this problem. Another method known as flash evaporation is often used in which pellets of the compound are dropped onto a "hot plate" where they are immediately vaporized.

In commercially available vacuum (or coating) units, provision is made for changing substrates so that a number of these can be coated during one pump-down. Provision is also made for changing sources so that a number of materials can be evaporated during one pump-down, for example, resistor material, dielectric material, conductors, and possibly semiconductors. Where out-of-contact masking is used (see § 8.3) provision is also made for mask changing, so that various patterns can be aligned in turn on the substrate.

The thickness of the film must be measured, or monitored, as it is being deposited. With resistors, it is easy to monitor the resistance directly, and so the actual thickness is of secondary importance. For other types of films, the most popular way of monitoring thickness is to use a quartz crystal monitor. In this, a quartz crystal, which is the frequency controlling element of an oscillator, is mounted inside the vacuum chamber, and as material is evaporated it deposits on the crystal, thus increasing its mass and reducing its natural frequency of oscillation. The change in frequency of the oscillator can be calibrated directly in film thickness (each material having its own calibration factor).

There are many more details which require careful attention for the vacuum evaporation method to be successful; cleanliness of the system, including a consideration of vapour pressures of materials inside the vacuum chamber; type of vacuum pumps used; methods of achieving precise mechanical displacement of masks and substrates while retaining a high vacuum system; material used for heater sources and possible contamination from this, or actual alloying with the material being evaporated.

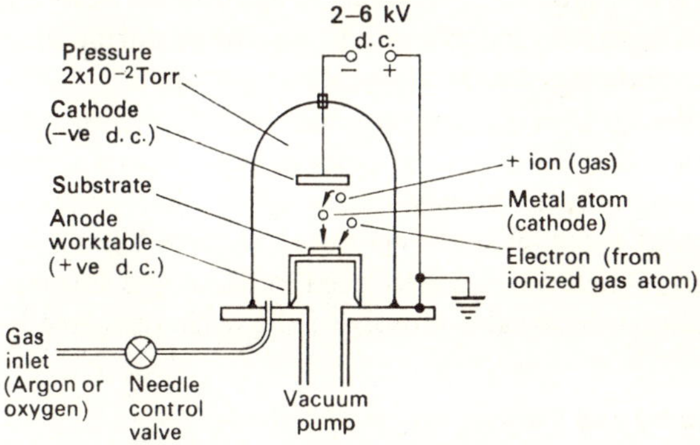

FIG. 8.2. A diode sputtering system.

Sputtering Methods

Figure 8.2 shows a basic diode sputtering system, so called because it contains two electrodes. One of the electrodes, the cathode, is made of the material which it is desired to deposit as a thin film, and this is held at a high negative potential (—2 to —6 kV) with respect to the anode during sputtering. The substrate is placed on the anode, which acts as a worktable. The bell-jar is pumped out as in the vacuum method, but then an inert gas (such as argon) is bled in through the needle valve, until the pressure rises to between 0.02 and 0.05 torr. With the application of the high voltage between the electrodes, the argon gas is ionized, and the positive ions are accelerated to the cathode. On striking the cathode, these liberate atoms of the cathodic material which diffuse through the space between cathode and anode, eventually coating the anode, including the substrate.

If, instead of argon, oxygen gas is used, oxygen ions will form an oxide with the cathodic material as it is liberated, and thus it is possible to deposit oxides in this way. The method is referred to as reactive sputtering (because of the reaction that takes place in the formation of the oxide). Another method of sputtering oxides is to use a cathode made of the oxide (and argon gas as before), but with this method it is not possible to employ the d.c. discharge because a positive surface charge collects on the cathode which repels further argon gas ions. Instead, a radio-frequency voltage is applied between cathode and anode, which permits the cathode to discharge during alternate half-cycles. The method is referred to as RF sputtering.

As with the vacuum evaporation method, it is the care and attention paid to details that determine whether or not successful thin films will be made. Here, only the briefest outline of the diode sputtering arrangement has been given. A more elaborate method known as the triode (or low-energy method) is widely used in which the substrate is not in the main discharge stream. However, the basic principle is the same.

8.3. Masking and Pattern-making Methods

The desired thin-film pattern (as distinct from the thin-film layer) on the substrate may be obtained by a number of methods. A widely used method in the vacuum evaporation process is to use a stencil-type mask in front of the substrate, the mask usually being made of thin metal foil. Although this method is very flexible, allowing a number of stencil masks to be aligned in turn and thus a number of patterns to be superimposed one on the other, the disadvantage is that the finest line-width that can be achieved is about 50 μm. Anything finer than this has to be achieved by highly precise and complicated mask displacement. The stencil mask, is held as close as possible to the substrate (to achieve the best possible definition) but can never be intimately in contact with it, and the method is therefore referred to as *out-of-contact* masking. Figure 8.3 shows a possible sequence of masks used in fabricating the simple circuit of Fig. 8.3f: (a) is the resistor mask; (b) the capacitor lower-electrode mask; (c) the dielectric mask; (d) the capacitor common-electrode mask; (e) the circuit as it appears on the substrate; and (f) the actual circuit.

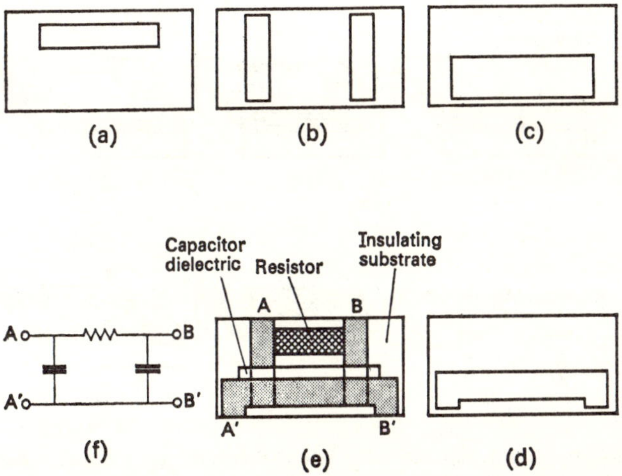

FIG. 8.3. Out-of-contact masking: (a) the resistor mask; (b) the capacitor bottom-electrode mask; (c) the dielectric mask; (d) the capacitor common-electrode mask; (e) the thin-film circuit on the substrate; (f) the circuit diagram.

The stencil masks may be fabricated using photo-resist methods in a manner similar to that described for the oxide masking for integrated circuits (Chapter 2). A major difference is that the line definition need not be so high, and therefore one stage only of photographic reduction usually suffices (about 20 : 1). The original artwork may be cut out on a co-ordinatograph, as described in § 2.2, or it may simply be laid out using precision tapes the widths of which are the desired line widths at the required magnification.

As an alternative to using a stencil mask, the substrate may first be completely coated with the required material, and this in turn coated with photo-resist. The photo-resist is exposed to ultraviolet light as shown in Fig. 2.3, which is then developed as described in § 2.3. Again, the difference is that the photographic plate need not have the definition required for integrated circuits if it is only being used to mask a single thin-film circuit. (Where, of course, thin-

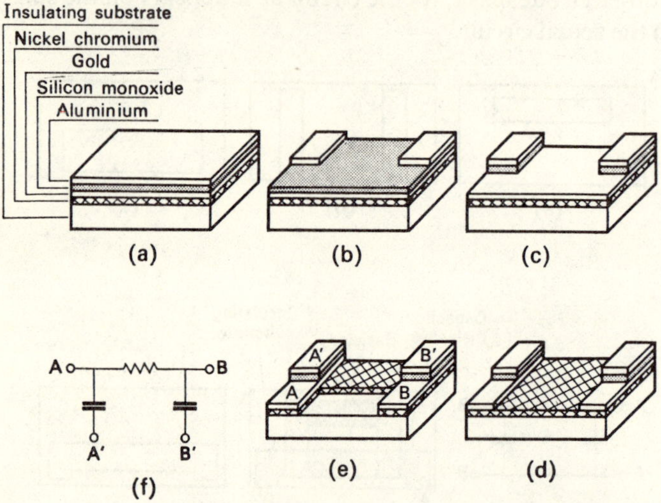

FIG. 8.4. In-contact masking using selective etching: (a) the uncommitted substrate carrying the various layers; (b) the capacitor top electrodes formed; (c) the capacitor dielectric formed; (d) the capacitor lower electrodes formed; (e) the resistor formed; (f) the circuit diagram.

film interconnecting patterns are being formed on integrated circuits, the fine definition is required.) Because the mask (the photo-resist) in this method is intimately in contact with the thin-film material it is known as *in-contact* masking. A very useful variation of in-contact masking known as selective etching, is illustrated in Fig. 8.4. Successive layers of material are deposited as shown in Fig. 8.4a (where thicknesses are greatly exaggerated for clarity): a selective etchant, which attacks aluminium only, is applied through a photo-resist mask to form the top capacitor electrodes (Fig. 8.4b); after cleaning, a new photo-resist mask is applied, and the next layer, silicon-oxide is etched, to form the capacitor dielectric (Fig. 8.4c); further cleaning and etching, this time the gold, forms the lower capacitor electrodes and connections to the resistor (Fig. 8.4d); finally, masking and selective etching of the nickel-chrome layer produces the resistor (Fig. 8.4e); the simple circuit chosen for illustration is shown in Fig. 8.4f.

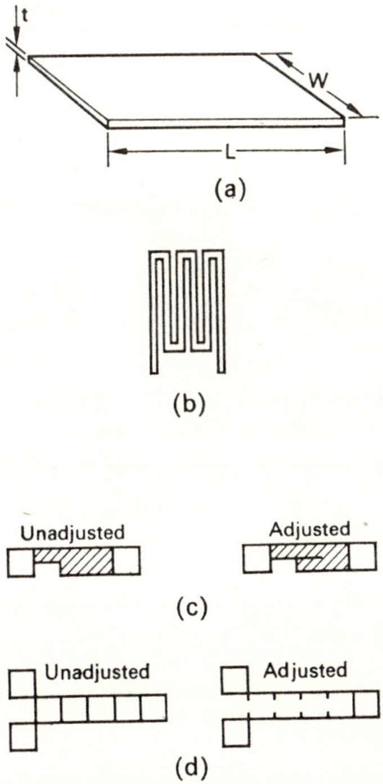

FIG. 8.5. Thin-film resistors: (a) a resistive sheet; (b) a meandered resistor; (c) and (d) methods of adjusting resistor values. ((c) and (d) Courtesy: Plessey Co. Ltd.)

Other methods which are commonly in use both for pattern making and for trimming component values (see later) come under the general heading of *micro-machining*. Here the thin film is physically cut using one of many possible tools. These range from a diamond scribe, to spark erosion, and laser- and electron-beam machining.

8.4. Thin-film Resistors

The simplest form of thin-film resistor is a sheet of material as shown in Fig. 8.5a. As with diffused resistors (see § 3.4) it is best to work in terms of sheet resistivity for the material. The resistance of the thin sheet of width W, thickness t and length L, with material of resistivity ρ (ohm-metres), when the dimensions are expressed in metres, is:

$$R = \rho \frac{L}{t\,W}, \tag{8.1}$$

and as before, the ratio ρ/t is defined as the sheet resistivity, ohms per square. Calling this R_s, the thin-film resistance is seen to be:

$$R = R_s . a, \tag{8.2}$$

where a is the aspect ratio L/W.

TABLE 8.1.

THIN-FILM RESISTOR MATERIALS (COURTESY: MOTOROLA SEMICONDUCTORS)

Material	R_s (ohms/sq)	Temperature coefficient (p.p.m./°C)	Deposition tolerance (%)
Nichrome	10–400	−100– +100	5
Tin oxide	25–1000	−500– +500	15
Tantalum nitride	50–500	−100– +100	10
Tantalum–chromium–silicides (Cermets)	100–20000	−300– +300	20

Where it becomes necessary to increase the aspect ratio, in order to achieve a higher resistance value, the resistor path length may be meandered as shown in Fig. 8.5b, and for very high-value resistors, a material having a high sheet resistivity must be used. Table 8.1 lists some of the materials commonly used, showing the range of sheet resistivities obtainable (by varying thickness for a given material), and also the important parameters of deposition tolerance, and temperature

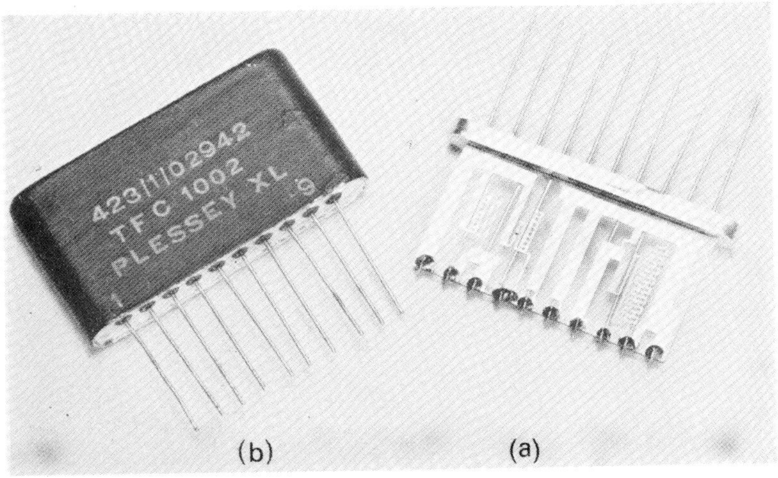

FIG. 8.7. (a) A thin-film circuit mounted on its connector strip; (b) the circuit packaged: (Courtesy: Plessey Co. Ltd.)

coefficient. The latter is usually measured in parts per million per degree centigrade, abbreviated as p.p.m./°C. Regarding resistor tolerance, this can be improved over the deposition tolerance by trimming the resistor as shown in Fig. 8.5.

Another important factor is the power dissipation and this is typically of the order of 2–3 W/cm^2 of resistor area.

In order to produce resistors that are stable with time, the deposited resistor is usually heated for a period, which tends to oxidize the surface and forms a protective layer, and during this baking cycle the resistance value tends to stabilize. To obtain a percentage tolerance better than the deposition value, the oxide-type resistors may be "trimmed" by further oxidizing until the required high tolerance resistance value is reached. With the nichrome resistors, some form of mechanical cutting is necessary, as shown in Figs. 8.5c and d.

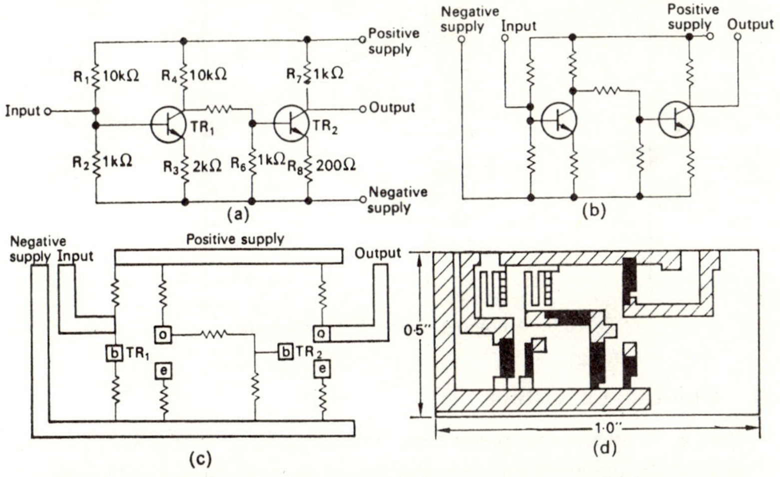

FIG. 8.6. Stages in the design and layout of a thin-film circuit: (a) the circuit diagram; (b) the substrate layout; (c) position of connector pads; (d) the final thin-film circuit (note that the transistors must be "wired" in as discrete components). (Courtesy: Plessey Co. Ltd.)

Figure 8.6 shows the stages in the design and layout of a typical thin-film circuit. It will be noticed that some of the low aspect-ratio resistors have notches, and these enable a diamond scribing tool to be accurately positioned for trimming; also shown are two ladder-type resistors.

Figure 8.7a shows a precision thin-film resistive network prior to encapsulation, and Fig. 8.7b the packaged circuit.

The nomograph presented in Fig. 8.8. enables a quick estimate to be obtained for values of *R*, R_s, and *a*.

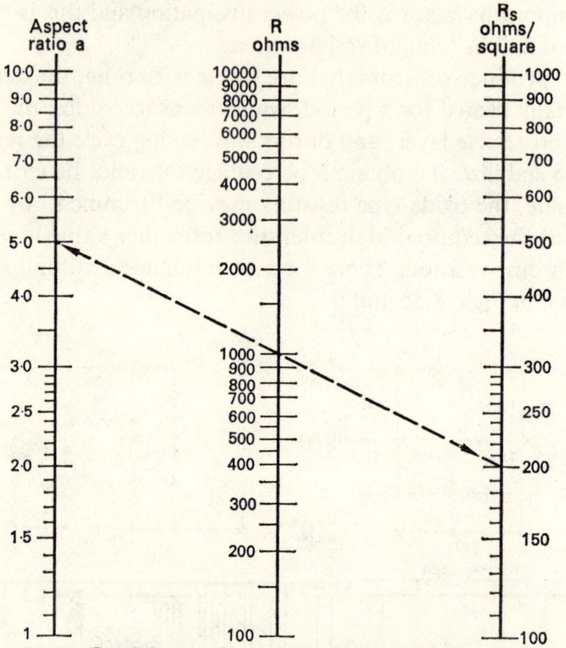

FIG. 8.8. A nomograph for film-type resistors.

8.5. Thin-film Capacitors

Figure 8.9a shows the basic thin-film capacitor, and it will be observed how the electrodes overlap at the edges to ensure a well-defined area for the plates. In order to be able to adjust capacitor values, spurs may be added, as shown in Fig. 8.9b, and it then becomes necessary to trim the spurs (i.e. cut them out as required) in a manner similar to trimming resistors. Since capacitance has to be removed in discrete quantities the adjustment is not as precise as that for resistors, and in circuits where the *C–R* time-constant is the important parameter, it is worthwhile trading capacitor tolerance for resistance tolerance. Table

TABLE 8.2.
THIN-FILM DIELECTRICS

Material	Dielectric constant	Temperature Coefficient (p.p.m./°C)	Capacitance pF/mm^2 for 1000 Å thick
Silicon monoxide	6	+150 to +250	531
Silicon dioxide	4	+200	354
Aluminium oxide	9	+300	796
Hafnium dioxide	40–80	+125	3500–7000
Tantalum pentoxide	20–25	+150 to +300	1770–2212

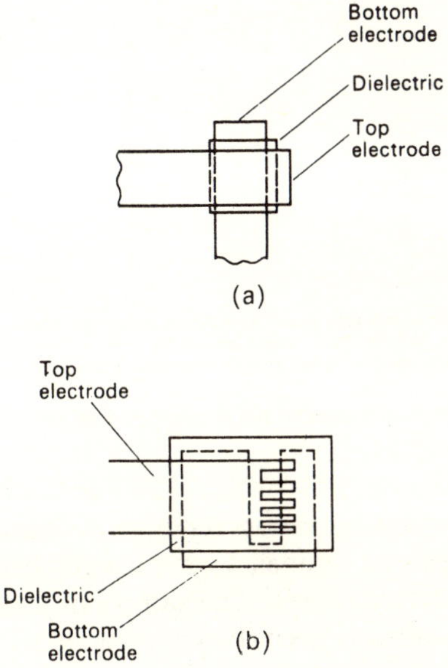

FIG. 8.9. Thin-film capacitors: (a) a simple capacitor; (b) a capacitor with provision for adjustment.

8.2 lists some of the materials that may be used for thin-film capacitor dielectrics. The important limitation on thinness is dielectric breakdown, and dielectrics thinner than about 200 Å cannot be used because of quantum-mechanical tunnelling which occurs, causing currents to flow through the dielectric.

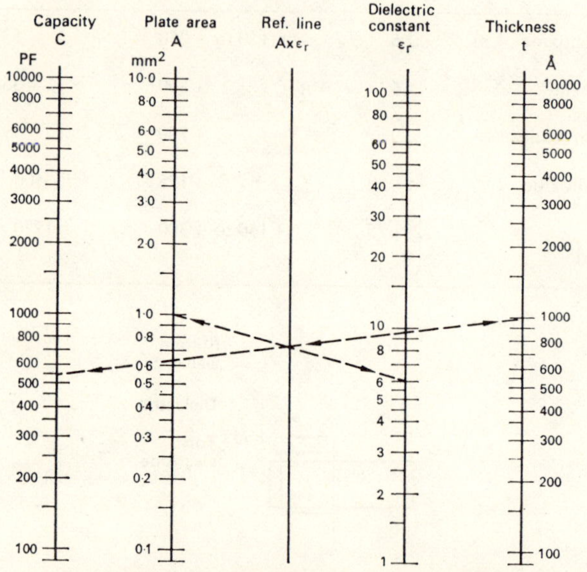

Use st. edge to join area A to Dielectric constant ε_r. Mark intercept on $A x \varepsilon_r$ ref. line. Join mark on ref. line to thickness t and project using st. edge to read required capacitance C.

FIG. 8.10. A nomograph for film-type capacitors.

The *Q*-factor of thin-film capacitors tends to be low (<200) and to drop rapidly with increase in frequency, this usually being due to the resistive losses in the electrode leads. An example of a circuit utilizing thin-film capacitors has already been shown in Fig. 8.3.

Figure 8.10 shows a nomograph for the calculation of thin-film capacitance.

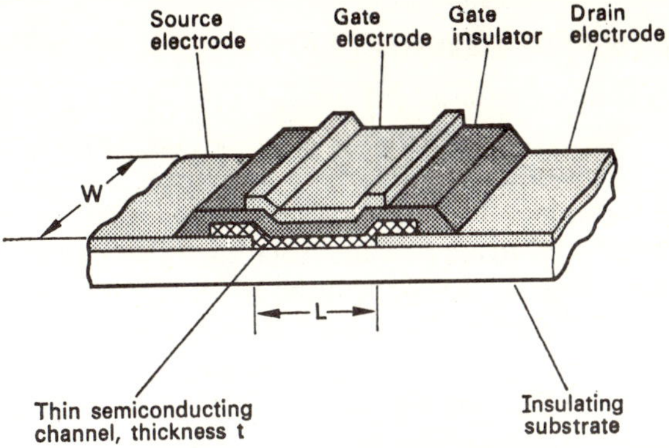

FIG. 8.11. The thin-film transistor.

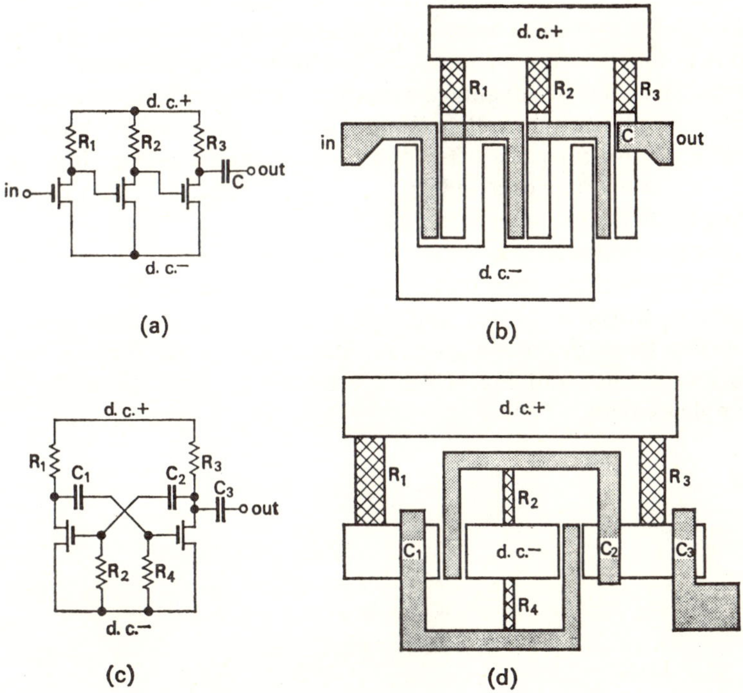

FIG. 8.12. Integrated thin-film circuits: (a) direct coupled amplifier; (b) the thin-film layout for (a); (c) a multivibrator; (d) the thin-film layout for (c).

8.6. The Thin-film Transistor

The thin-film transistor is very similar to the MOST already described (Chapter 6) the essential difference being that the conducting channel is formed by means of a thin film of semiconductor material. Figure 8.11 shows one form the thin-film transistor may take. It will be observed that the device is fabricated on an insulating substrate, which is one of its advantages over presently produced MOSTs. The source and drain contacts at either end of the semiconducting channel are ohmic, i.e. ideally there should be no *p–n*-type junction at these contacts, and in this respect the TFT differs also from the MOST, where it was seen that the source and drain contacts were formed as *p–n* junctions with the substrate.

The channel conducting mechanism, and the method of field control through the capacitive gate, is identical for the two devices (TFT and MOST) and therefore the equations developed in Chapters 6 and 7 apply to the TFT. One disadvantage with the TFT is that the semiconductor film is usually polycrystalline (see § 1.10), and this tends to reduce the field effect mobility. To offset this disadvantage, however, it is quite easy to use any one of a wide range of semiconductors and this gives great flexibility in design of circuits.

The two simple circuits shown in Fig. 8.12 illustrate the compatibility of the TFT with the other thin-film components. In Fig. 8.12a is shown a direct-coupled amplifier, and in Fig. 8.12b, one possible layout in thin-film form for the complete amplifier. Figure 8.12c shows a multivibrator circuit, and Fig. 8.12d a possible thin-film layout for the complete circuit.

CHAPTER 9

Thick-film Circuits

9.1. Introduction

For many applications the precision afforded by thin-film methods is not required. In the thick-film process, the circuit patterns, including components, are printed and fired onto the substrate. The result is a process which is readily automated and considerably cheaper than the corresponding thin-film process. Usually, thick-film circuits have the added advantage of being able to dissipate higher powers than the thin-film counterpart. As already mentioned in the previous chapter, thick films range from about 10 μm upwards (a typical upper limit being about 50 μm), while films just below the 10 μm limit may be classed as either thick or thin, depending on the method of fabrication.

At the present time, thick-film circuits tend to follow the same pattern as thin films in circuit applications, with resistors, capacitors, and conductor tracks being the main components that are fabricated together. Inductors can be made, but these usually require plating to increase thickness, and the Q-factor still tends to be low. Experimental thick-film transistors have also been made on a laboratory scale, and these certainly offer attractive possibilities if the process can be adapted to production while maintaining reasonable yields.

9.2. The Silk-screen Process

The Silk-screen Process is the most common method of applying thick-film patterns onto a substrate. This is a process adapted from the printing industry, and consists essentially of printing the required pattern through a stencil screen. Although in the printing, or graphic arts industry, the screen is made of silk, in thick-film circuit applications, stainless steel is used, (or sometimes nylon) in order to maintain a high tolerance of registration of successive prints and good repeatability of patterns. The printing inks used in circuit applications must also have the required electrical characteristics, e.g. highly conducting for conductor patterns, resistive qualities for resistor components, etc.

The stencil pattern may be formed directly onto the screen by impregnating this with photo-resist, and developing the required pattern by the methods already described (§ 8.3). Although such a

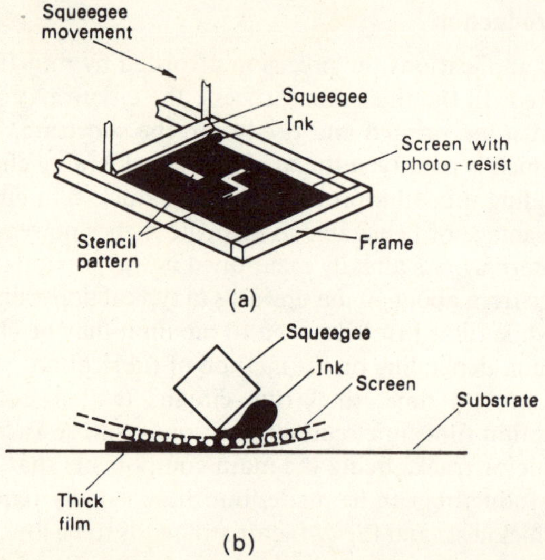

FIG. 9.1. The silk-screen process: (a) a sketch of the screen (usually stainless steel or nylon) carrying the desired pattern, and the squeegee; (b) an enlarged view of squeegee action.

stencil mask is mechanically strong, its accuracy is limited, and the alternative, known as the indirect method, is to make the mask on a separate sheet of material and attach this to the screen for support. Yet another method, where very fine lines are required, is to make metal (stainless steel or molybdenum) out-of-contact masks similar to those used in thin-film work (§ 8.3). Figure 9.1a illustrates the basic features of a thick-film printing system. The ink, having the desired characteristics, is forced through the exposed regions of mesh by the passage of the

FIG. 9.2. A thick-film resistive circuit. (Courtesy: CTS Corp., Elkhart, Indiana, U.S.A., and A.B Elect. Comp., Rhondda, Glam.)

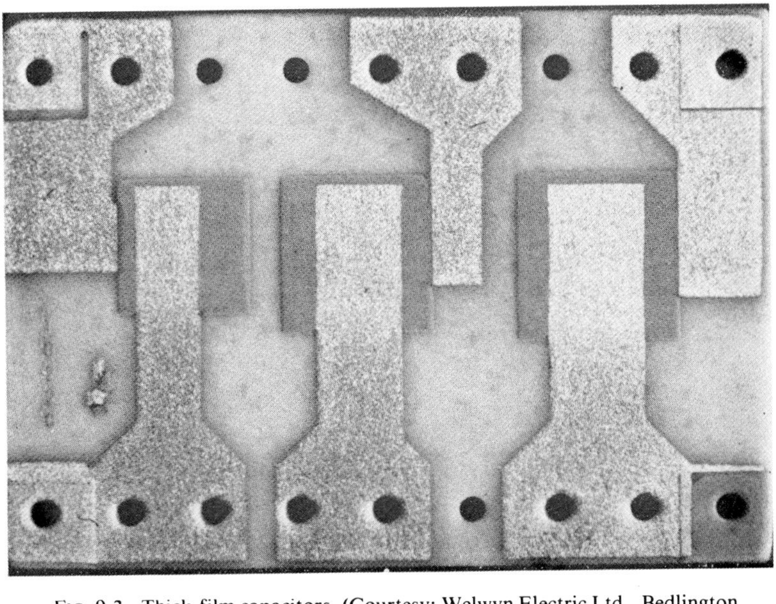

FIG. 9.3. Thick-film capacitors. (Courtesy: Welwyn Electric Ltd., Bedlington, England.)

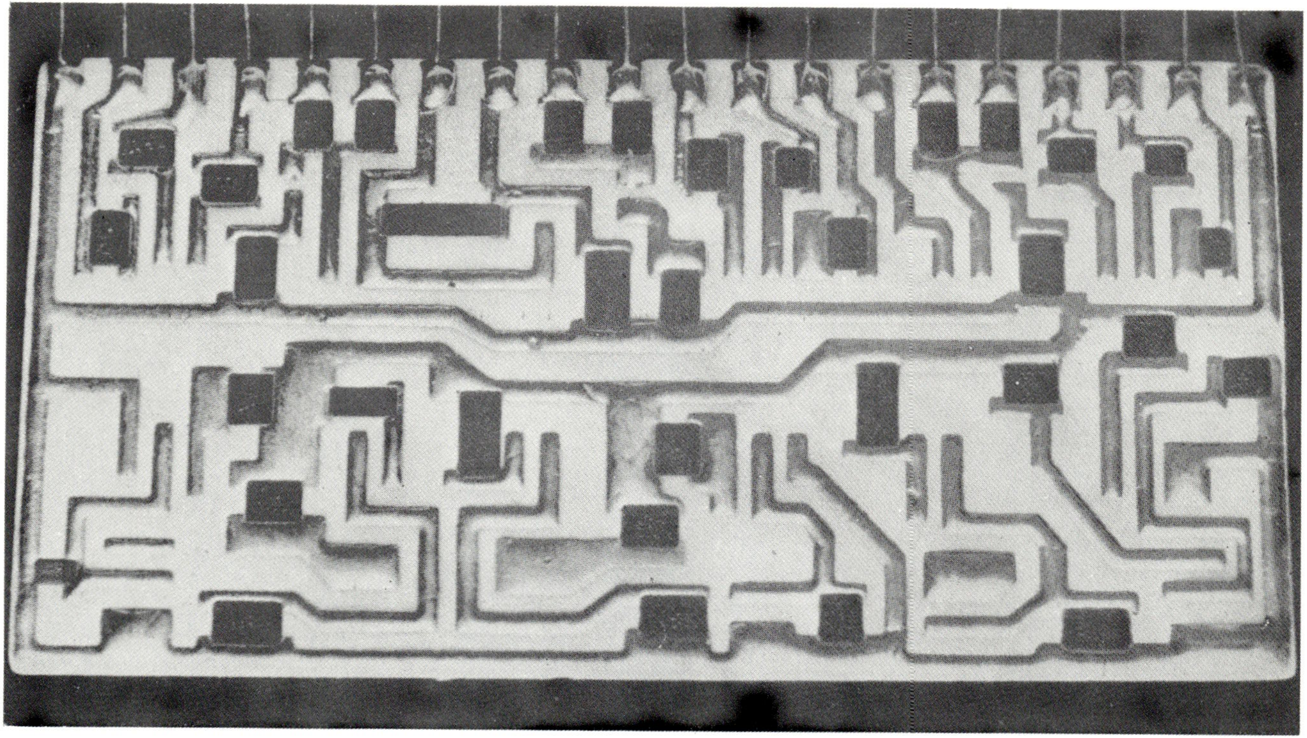

FIG. 9.4. Two- by three-inch thick-film interconnecting pattern incorporating close tolerance, high dissipation thick-film resistors. The assembly is intended for use in a hybrid circuit with silicon integrated circuits. (Courtesy: ITT Components Group, Europe.)

squeegee over the screen, and usually, a continuous feed system for the ink is employed which just replenishes the ink being used. The ink must be evenly spread along the length of the squeegee. Figure 9.1b illustrates how the ink flows through screen mesh.

Although the process sounds comparatively simple, there are many factors which have to be accurately controlled in order to obtain reproducible results. Some of these are: the pressure on the squeegee, the velocity of the squeegee travel, the snap-back distance (i.e. the distance from screen to substrate without squeegee pressure), the angle of squeegee blade with screen. For best results the squeegee should only be moved in one direction for printing. The viscosity of the ink is also important, as it determines the rate of ink flow through the mesh, and, of course, the mesh size will determine to a large extent the line definition obtainable (mesh sizes ranging from 165 to 325 holes per linear inch are in use).

Ink Composition

The inks are composed of (i) an organic resinous filler, which acts as the screening vehicle, i.e. it carries the other components of the ink through the mesh; (ii) a solvent, which is added in desired quantities to control the viscosity of the ink; (iii) a low melting-point, highly reactive glass frit, which provides bonding between circuit and substrate in the final process; (iv) a metal oxide where capacitor dielectrics are being formed, and a metal for resistors or conductor patterns (obviously the resistivity of the metal and the particle density will depend on the application).

Following the printing of the circuit and components, the assembly is dried at about 100–150°C to remove the solvent and then passed through a furnace (i.e. fired) at about 760–1000°C, to decompose the organic filler. The firing also sinters the glass frit particles into the substrate and provides the bonding between the circuit and substrate. Control of the firing temperature is highly critical, and usually the furnace has heating zones controlled to ±1°C, and cooling zones controlled to within ±2°C. The substrates are carried through on a belt-feed system, a typical speed being 4 in./min, and a typical total furnace length being 10 ft. Provision is also made for introducing various gas ambients into the furnace tube.

Care must be taken when a number of patterns have to be processed sequentially, as later firing stages may affect the previously deposited patterns. This is particularly true of capacitor dielectrics where the binder component (the glass frit) for the bottom electrode may melt and diffuse into the dielectric thus affecting the dielectric constant.

9.3. Resistors

A typical resistor material is palladium silver–palladium oxide composition (PdAg/PdO). Sheet resistivities in the range 50 Ω to 15 kΩ per square can be achieved. Power dissipation ranges between 6 W/cm^2 and 10 W/cm^2, and typical resistors show less than ½ per cent drift after 3500 hr life test. The temperature coefficient of resistance is of the order of +100 p.p.m./°C, but depends to some extent on the temperature range. Figure 9.2 shows a simple thick-film resistive circuit.

9.4. Capacitors

Most dielectrics in use are proprietary compositions of oxides, but typically, for low-frequency applications they may contain barium titanate (BaTiO), while for high frequency (e.g. R.F.) titanium dioxide (TiO_2) is used. With present methods of manufacture, dielectrics are usually made thicker than 50 μm, as it has been observed that dielectrics thinner than this show a drop in insulation resistance when subjected to a 10 V, 500–700 hr life stress test. The reason for the drop may be ion drift in the dielectric, but this has not been established.

Thick-film capacitors show a drift with temperature which tends to be rather "peaky". For example, from —55° to +25°C, a typical drift figure is —15 per cent, and from 25° to 85°C, + 7 per cent. Also, most dielectrics show a hysteresis effect, that is, capacitance changes with applied voltage, and the change is different for decreasing voltage compared with increasing voltage.

Capacitance and loss factor variation with frequency is negligible up to frequencies of about 10 MHz, which is good compared with thin-film capacitors, and a typical loss factor is 2 per cent. Figure 9.3 shows a simple thick-film capacitor assembly.

9.5. Conductor Patterns and Substrates

Precious metals are normally used for conductors, because they have low resistivity, and also because they do not oxidise. Typical sheet resistivities for conductors are less than 0.01 Ω per square. Although gold is suitable as a low resistivity material, it adheres weakly to the substrate, and as a result, gold platinum alloys are used (Au/Pt). The smaller grain size of the alloy forms a stronger bond with the glass-frit than gold alone. Figure 9.4 shows a thick-film interconnecting pattern to which integrated circuits will be attached separately to form a hybrid circuit (see chapter 10.) The thick-film interconnecting pattern also incorporates thick-film, close tolerance, high dissipation resistors.

The substrate material for film circuits (thick or thin), must be carefully selected with a number of points in mind. It must have good electrical insulation; its temperature coefficient of expansion must not be too great, and should match as closely as possible that of the capacitor dielectrics; thermal conductivity should be high; other factors which usually have to be closely controlled are surface finish and surface flatness. A typical substrate material is alumina (Al_2O_3), which is 95 per cent pure.

CHAPTER 10

Hybrid Circuits

10.1. Introduction

During the early days of development in microelectronics, silicon integrated circuits and thin-film circuits were often considered as competitors to one another. For the large majority of applications, however, there is a need for combined circuitry, utilizing silicon active devices and film-type passive components and interconnections (with the film circuits now including thick films). These circuits, known as *hybrid circuits,* have an assured future in electronic circuit applications if only because they exploit the best advantages of both types of circuits.

Hybrid circuits are especially useful with linear, or analog, circuits, where the variety of circuits required, and the rather more demanding tolerances on components (compared with digital circuits) would make completely integrated circuits very costly. Another major difficulty with linear integrated circuits is that of obtaining inductance. Generally it will be found that for tuned amplifiers, or where frequency selection is required, the frequency determining components are added externally either in film form or as discrete components. At the research and development level, considerable effort is being directed into the design of frequency selective circuits that do not require inductors. However, in this chapter, some of the methods of forming hybrid circuits, useful both for connecting integrated digital circuits into larger arrays, and for overcoming the shortcomings of linear integrated circuits will be examined. (It should be noted that in a sense, all integrated circuits are hybrid circuits, since the metallizing interconnecting layer is deposited by thin-film means.)

10.2. Hybrid Assembly on a Chip

Figure 10.1 shows a 60 MHz amplifier which is constructed using a silicon integrated circuit for all the active devices, and thin-film resistors and capacitors. The main reason for using thin-film resistors here is that the capacitance associated with diffused-type resistors (see § 3.4) would be a severe limitation on the gain-bandwidth. Nichrome film of 300 Ω per square is used for the resistors. Thin-film capacitors are also used to avoid the parasitics encountered with diffused capacitors, and for the capacitors shown, 460 pF/mm^2 is achieved. Figure 10.1a shows the basic amplifier circuit, which is a cascode amplifier; Fig. 10.1b shows the initial layout; and Fig. 10.1c shows the single silicon die with the passive thin-film components deposited on the top surface. It will be noticed that the inductively coupled circuits which provide the input and output coupling must be added separately, as discrete components (and they do not appear on the circuit die, Fig. 10.1c). These will be conventional miniature IF transformers, untuned.

10.3. Hybrid Bonding Methods and Circuits

Ultrasonic bonding

Figure 10.2 shows, in a simplified manner, the sequence of events that occur in bonding a wire between a film-circuit and a silicon chip, using ultrasonic means. In the ultrasonic method, the wire is held under pressure, and ultrasonically vibrated on the region to which it must be bonded. The intense and very localized vibrations cause the wire to weld onto the surface. The advantages of the ultrasonic method are that any heat generated is confined to the small region around the bond, and also, bonds can be made between the wire and different types of surfaces, including the insulating substrate. (The latter may be used for anchoring a wire.)

The welding tip has to be made of very hard material (e.g. tungsten carbide), and Fig. 10.2a shows the tip and wire being brought into contact with the silicon chip. The positioning of the chip under the tip has to be accurate, and is achieved through a micro-positioning mechanism and microscopic viewing arrangement. When the wire has been

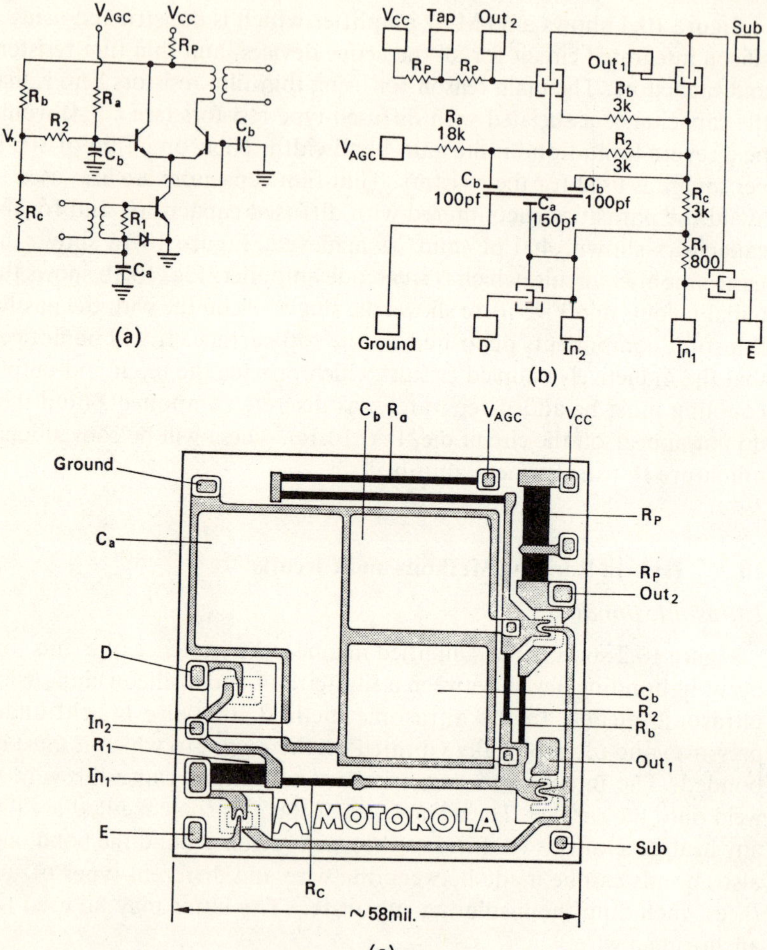

FIG. 10.1. A hybrid-type integrated circuit utilizing silicon-integrated active components and thin-film passive components: (a) the basic cascode circuit for a 60 MHz I.F. amplifier; (b) the initial layout; (c) the silicon die containing the circuit. Note that the inductive components must be added separately. (Courtesy: Motorola Inc.)

bonded onto the chip, the tip is lifted and the substrate moved by means of the micro-positioner so that the film circuit is positioned under the tip (Fig. 10.2b). After the second bond is made a clamping mechanism breaks the wire within the tip, and when the tip is lifted the wire is

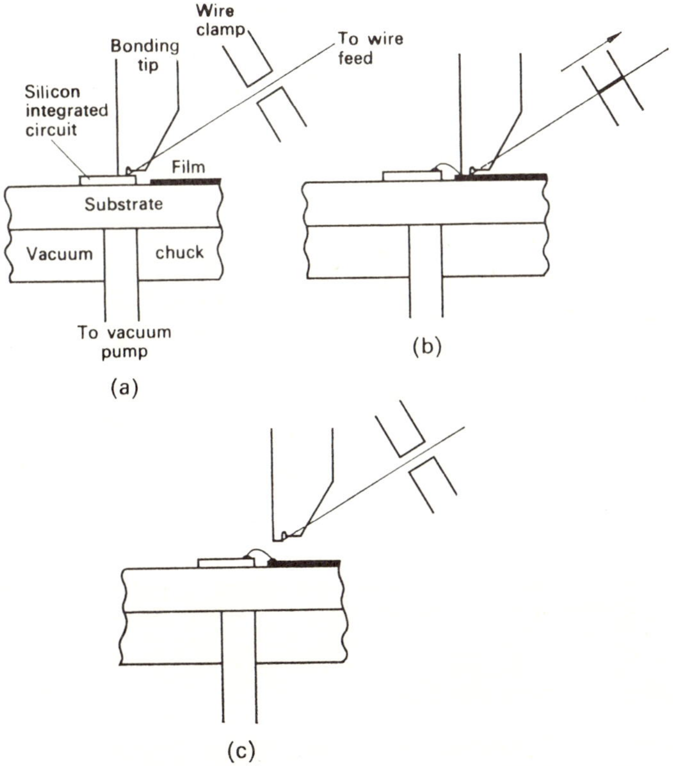

FIG. 10.2. Ultrasonic wire bonding: (a) the wire being bonded to the chip; (b) to the film circuit; (c) the finished bond.

pushed forward into position, ready for the next bonding operation (Fig. 10.2c). In commercially available bonders, two independent ultrasonic energy sources can be selected by means of a switch, as the bonding energy for wire on silicon will be different, in general, for that for wire on film. It should be noted that the chip itself may be ultrasonically bonded onto the substrate.

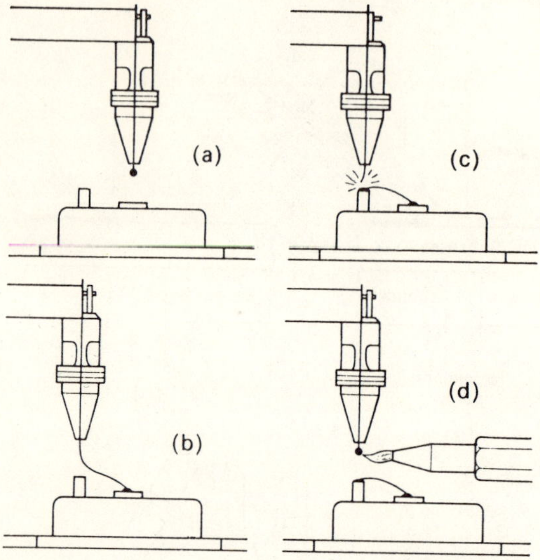

FIG. 10.3. Thermocompression bonding: (a) tip with ball-ended wire positioned for bonding; (b) silicon bond formed; (c) post bond formed; (d) ball-end formed ready for next bond. (Courtesy: Axion Corp.)

Thermocompression Bonding

In this method, illustrated in Fig. 10.3, the bonding tip is heated, and brought down, with the wire, onto the surface to which the bond is required. Figure 10.3 shows the sequence of events in bonding a wire between a silicon chip and a post in a TO-5 type can, but of course, the same sequence could be used to bond to film circuits. In Fig. 10.3a, the silicon chip is positioned under the wire, and it will be noticed that the wire has a ball formed at the end. The tip is electrically heated, and brought down onto the chip, to form the thermal-compression bond, as shown in Fig. 10.3b. The work is repositioned so that the post is under the tip, and a similar bonding operation occurs, except this time, a wedge bond, rather than a ball bond, is made. A clutch clamps the wire, and breaks this off as the tip is raised (Fig. 10.3c). Finally, a hydrogen flame is passed across the tip of the wire protruding from the welding tip, to form a ball, in readiness for the next bonding operation (Fig. 10.3d).

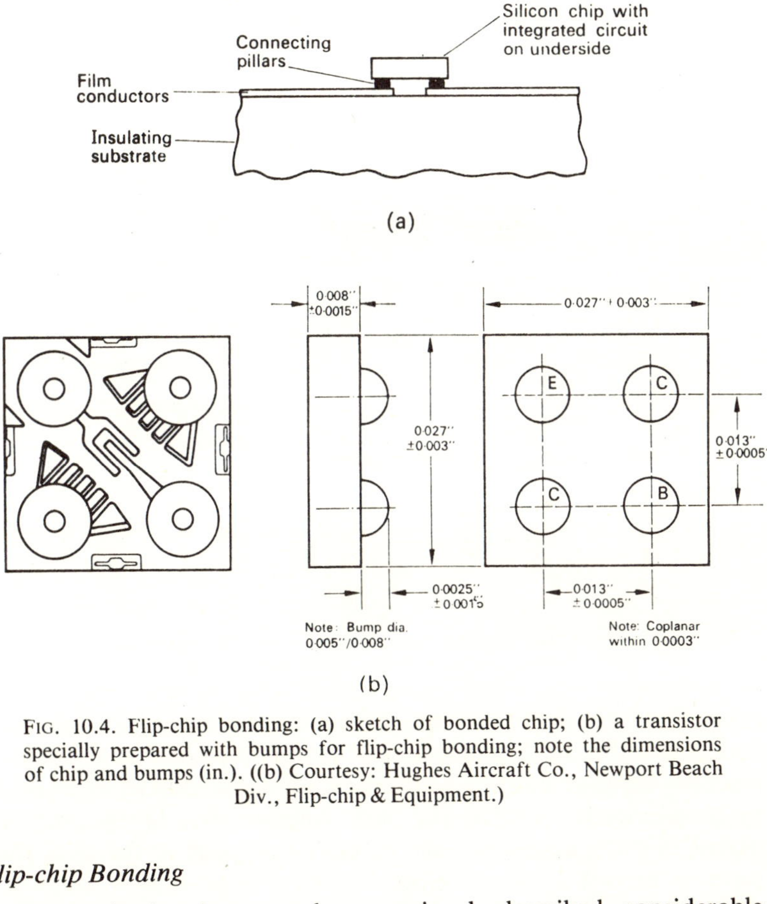

FIG. 10.4. Flip-chip bonding: (a) sketch of bonded chip; (b) a transistor specially prepared with bumps for flip-chip bonding; note the dimensions of chip and bumps (in.). ((b) Courtesy: Hughes Aircraft Co., Newport Beach Div., Flip-chip & Equipment.)

Flip-chip Bonding

Where wire bonds are used, as previously described, considerable space is wasted in the circuit as a whole. Also, the wires have stray inductance which may adversely affect the performance of a circuit, and there is the obvious danger that the wire itself may break, resulting in circuit failure. In an effort to overcome the disadvantages of wire bonding, various methods of bonding the silicon circuit directly into the film circuit have been proposed, of which the flip-chip method is one. As the name suggests, the silicon chip is flipped onto the substrate carrying the film circuit, and the electrical bond is made through specially prepared pillars or "bumps". These may be attached initially

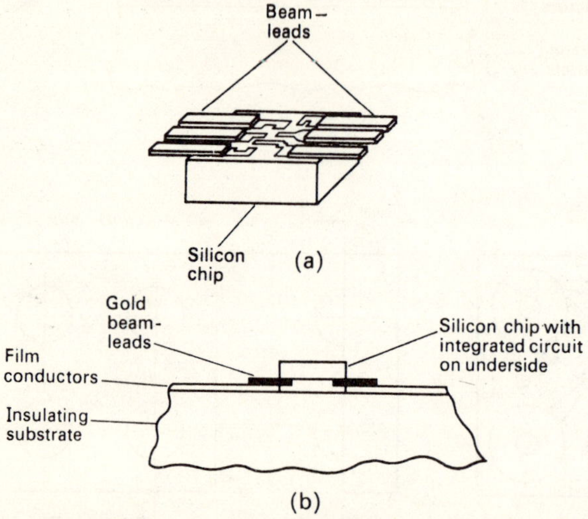

FIG. 10.5. Beam-lead bonding: (a) a chip prepared with beam leads; (b) the chip beam lead bonded onto a film circuit.

to either the chip or the substrate, or they may be inserted separate from both. Once the chip is flipped, and the bumps located at the desired positions between chip and film circuit, the actual bond may be made by ultrasonically vibrating the chip. Alternatively, hot gas may be passed over the assembly while pressure is applied to the chip, thus forming thermal-compression bonds. Figure 10.4a shows a sketch of a chip that has been flip-chip bonded, and Fig. 10.4b shows in plan view a transistor specially made for flip-chip assembly, in this case the bumps

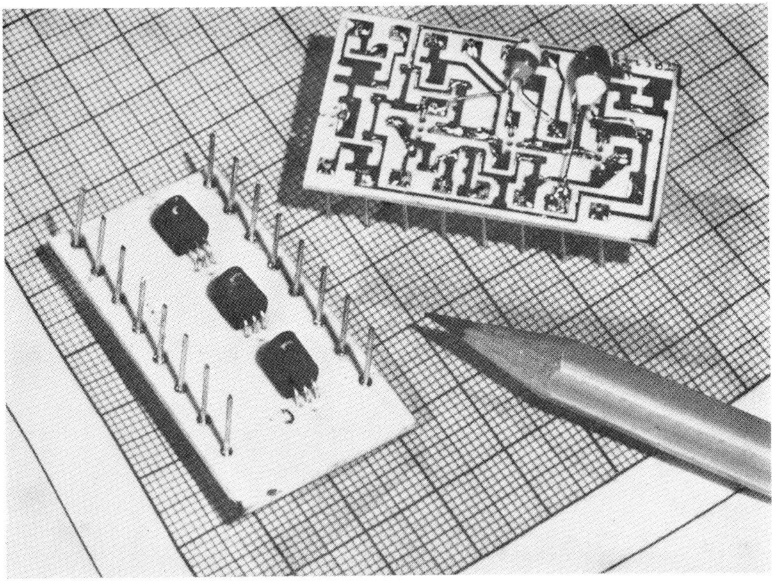

FIG. 10.6. A thick-film circuit with added transistors and miniature tantalum electrolytic capacitors. (Courtesy: ITT Components Group, Europe.)

FIG. 10.7. A thick-film circuit with added integrated circuit in a flat-pack. The thick-film resistors have been adjusted to ultra close tolerance by air-brasion. (Courtesy: ITT Components Group, Europe.)

being formed onto the transistor chip. Also shown are the dimensions (in inches) of the flip-chip transistor. A similar technique known as a "leadless inverted device" or lead is described in § 11.5.2.

Beam-lead Bonding

As an alternative to flip-chip bonding, special leads, known as *beam leads* (because they overhang the chip like rigid beams) are formed onto the chip, while it is an integral part of the larger silicon slice. Instead, then, of dicing the slice into the required chips, these are etched out, leaving the gold beams protruding over the edge, as shown in Fig. 10.5a. The chip, with the beam-leads, is then inverted onto the film circuit, and the necessary bonds made between the beam leads and film circuit (usually by thermal compression bonding), as shown in Fig. 10.5b. A beam-lead assembly for a microwave diode is shown in Fig. 11.12.

Re-flow Solder Method

With thick-film circuits, a normal flow-solder technique may be used to plate up the thickness of the conductors (see Fig. 9.4) and this also facilitates the addition of discrete components. By incorporating a hole pattern in the substrate, discrete components may be added on both sides. Figure 10.6 shows a thick film circuit carrying miniature tantalum electrolytic capacitors on the film side, and transistors on the underside. Figure 10.7 shows a thick-film circuit carrying a semiconductor flat-pack. The thick-film resistors shown have been adjusted to ultra close tolerance by air-brasion. In this process, which is extensively used for the adjustment of thick-film resistors, a jet of high-pressure air containing aluminia particles (size approximately 20 μm) is used to abrade away a section of the resistor track. The resulting increase in the resistor aspect ratio produces an increase in resistance value.

The soldered connections between component leads and the solder "lands" on the circuit are formed by remelting the solder with the components in position, this being termed *reflow soldering.* Reflow soldering is also used to connect silicon integrated circuits into printed circuit boards.

CHAPTER 11

Microwave Applications of Microelectronics

11.1. Introduction

In common with other integrated circuits, microwave integrated circuits hold out promise of higher reliability, greater functional versatility, smaller size, and lower cost than conventional microwave circuitry. Although the term "integrated circuit" is used here, almost all microwave integrated circuits are hybrid assemblies; that is, passive components such as resistors, transmission lines, etc., are made in film form, and the active devices are connected in as discrete components, by one or other of the bonding methods discussed in Chapter 10. Monolithic microwave circuits can be made (monolithic meaning in a single block or chip (of semiconductor)), but the main difficulties encountered are, (i) achieving low ohmic contacts, and (ii) conductivity modulation effects (this is similar to the effect discussed in connection with Fig. 6.8). It would seem, therefore, that for many years to come, hybrid circuits will be used at microwave frequencies, and it is quite possible that further developments will depend on utilizing completely new properties of semiconductors at microwave frequencies. Microwaves may be considered to cover the frequency range 1 GHz to 30 GHz (1 GHz = 10^9 Hz), or wavelengths from 30 cm to 1 cm.

11.2. Passive Components and Microstrip Transmission Lines

Passive components for use at microwaves may be arranged as lumped component circuits, or alternatively, they may be made distributed over a considerable area of substrate (when they are known as distributed circuits). The determining factor is the dimensions of the circuit compared with the wavelength in use. If a circuit length, for example, is of the same order as the wavelength, then the circuit is a distributed one.

For both types of circuits, the surface finish on the substrate must be optically flat; polished alumina (Al_2O_3) is a typical substrate material.

Lumped Components

Resistors, inductors, and capacitors, can all be made using either thin-film or thick-film techniques, although thin-film methods provide better control of circuit parameters. It will be noticed that inductors are included, the reason being that at microwaves, the inductance required for circuit functions is small (of the order of a few nanohenrys). Resistors are made using standard thin-film techniques, and the range most commonly required is 20 to 200 Ω. These can be made using nichrome, with simple rectangular geometries (i.e. non-meandered).

Capacitors, where comparatively high values are required (e.g. 10 pF), are made also as normal thin-film capacitors, with overlapping plates separated by the dielectric. Where smaller values of capacitance are required, an interdigitated-type construction is used (Fig. 11.1). Here, both plates of the capacitor lie in the same plane, and capacitances in the range 0.1 pF to 0.4 pF can be obtained with very high *Q*-factors.

Inductors in the range 0.5–2.0 nH, and with *Q*-factors of the order of 50 can be achieved using the construction shown in Fig. 11.2a. In constructing these, a 300 Å layer of nichrome is first evaporated onto the substrate, followed by a 0.5 μm, gold layer. This is plated up to a thickness in the range 3–5 μm, to improve the *Q*-factor. Figure 11.2b shows how inductance and *Q*-factor vary with width. Resonant circuits may be formed as shown in Fig. 11.3. Fig. 11.3a shows a series resonant circuit, and Fig. 11.3b a parallel resonant circuit.

The advantages claimed for lumped component circuits are that:

(i) They can be made small physically (in terms of wavelength).

(ii) They do not require high dielectric constant substrates because the circuits are deposited on one side of the substrate (compared with distributed circuits where a ground plane is used).

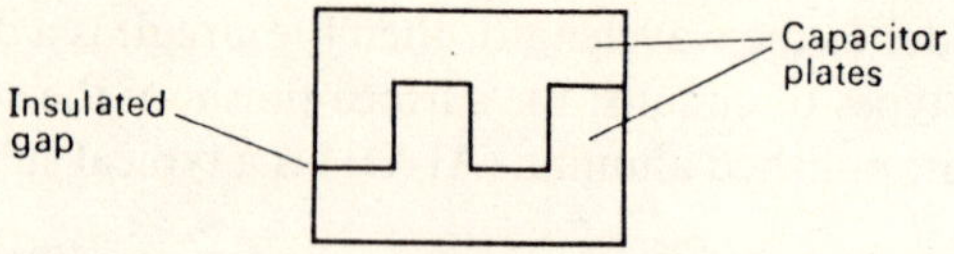

FIG. 11.1. A coplanar capacitor.

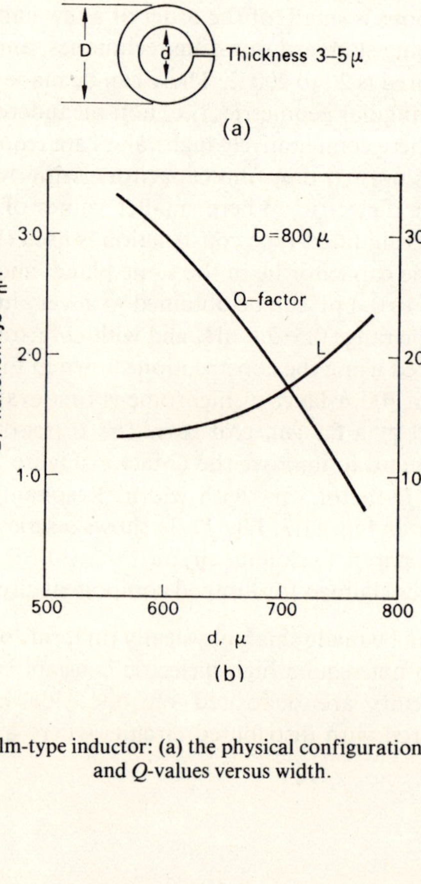

FIG. 11.2. A film-type inductor: (a) the physical configuration; (b) inductance and Q-values versus width.

(iii) Resonant circuits have only one resonant frequency compared with multiple harmonic resonances associated with distributed circuits (e.g. transmission lines).

Regarding the last point, it should be realized, however, that at sufficiently high frequencies, lumped circuits will eventually appear as distributed circuits, and care would have to be taken to ensure that parasitic resonances occur outside the useful frequency range. Also, in regard to point (i), the limiting factor in physical size may well be proximity effects of boundaries (e.g. screening cans) rather than the physical size of the circuit. Also, a metallic enclosure can behave as a resonant cavity which then requires the addition of absorbers to damp out unwanted resonances.

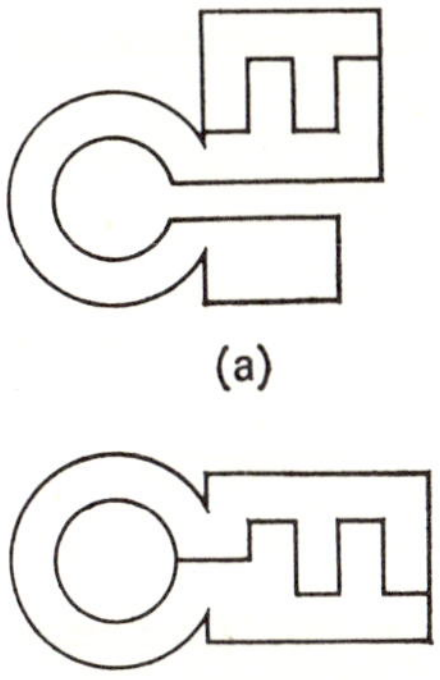

(a)

(b)

FIG. 11.3. Film-type resonant circuits: (a) series resonant; (b) parallel resonant.

Transmission Lines

The two most common forms of construction for transmission lines are the *microstrip line,* Fig. 11.4a, and the *triplate line,* Fig. 11.4b. Both types of line employ a ground plane (or ground planes) and a parallel signal conductor.

An important characteristic of transmission lines is the *characteristic impedance* Z_0, of the line, a knowledge of which is required, for example, to enable the line to be correctly matched to other parts of the circuit thus minimizing reflected waves. For the lines shown in Fig. 11.4 (and also in Fig. 11.5a), the characteristic impedance depends primarily on the ratio of conductor width b to line spacing h, and the dielectric constant ε_r of the substrate. Figure 11.5a shows how the product $Z_0\sqrt{\varepsilon_r}$ varies with the ratio b/h. The product $Z_0\sqrt{\varepsilon_r}$ is plotted rather than Z_0 in order to make the curves more general. If for example, the dielectric constant ε_r of the line is 5, then for $Z_0 = 50\ \Omega$ the vertical scale reading would be 112, and, from Fig. 11.5a, this would be achieved by a microstrip line for which $b/h = 1.8$, and by a triplate line for which $b/h = 0.82$.

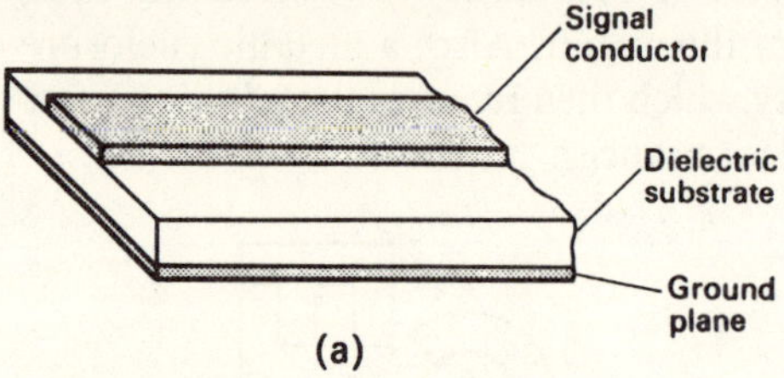

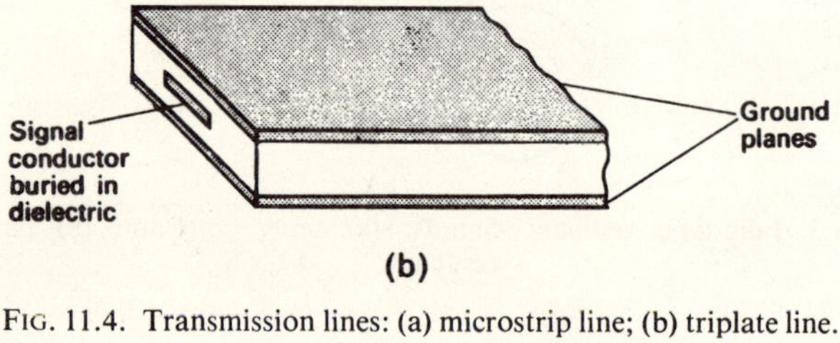

FIG. 11.4. Transmission lines: (a) microstrip line; (b) triplate line.

The results shown in Fig. 11.5a are obtained using a number of simplifying assumptions and should therefore only be used as a guide to show the approximate variation of Z_0 with line parameters. The main simplifications made are: (i) conductor thicknesses are neglected; (ii) only the dielectric constant of the substrate is taken into account; (iii) the ground plane is assumed to extend to infinity. For the microstrip line in particular, the fact that the total dielectric consists of both

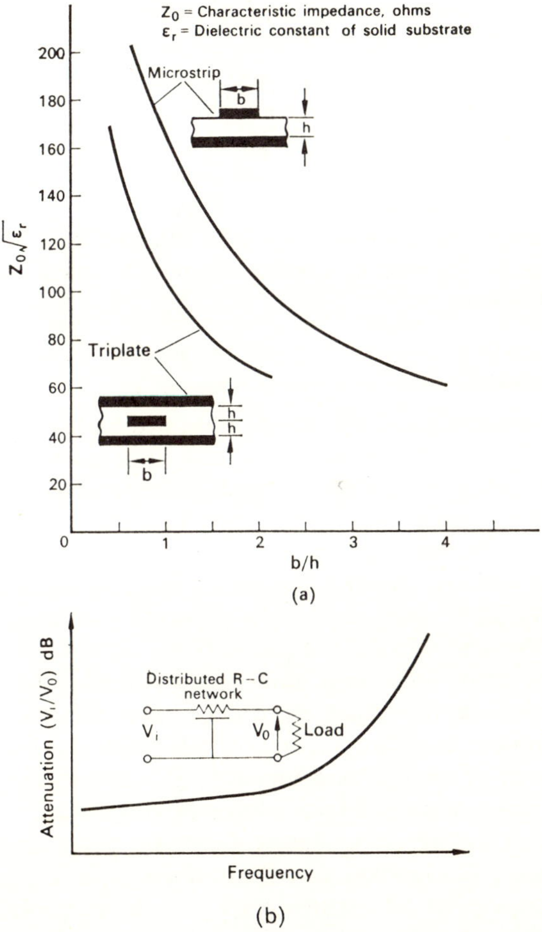

FIG. 11.5. (a) Characteristic impedance as a function of line dimensions. (b) A thin-film distributed network having the characteristics of a low-pass filter.

air and substrate introduces multiple modes of propagation, i.e. the line behaves partly as a normal transmission line and partly as a waveguide.

In practice, the infinite ground-plane is approximated very closely when the actual ground-plane width is greater than three times the conductor strip width. This also suggests that a spacing of at least three times the signal conductor width is required between signal conductors using the same ground plane, if interference (cross-talk) between circuits is to be avoided. As with lumped circuits, the substrate must be smooth, and polished alumina is commonly used.

There is a great variety of other circuits that can be made in distributed form (applicable to both microwave and other frequencies), and it is not possible to go into these here. However, by way of example, Fig. 11.5b shows the shape of a low-pass filter attenuation characteristic that can be achieved by making the microstrip line into a distributed R–C network.

11.3. Microwave Solid-state Power Sources

Although microwave power may be generated by fairly conventional means using transistors, the transistors have to be specially constructed in order to operate satisfactorily at the high frequencies involved. The upper useful frequency limit using transistors is about 3 GHz. Another fairly conventional method is to use varactors. (The word varactor is coined from *var*iable re*actor*.) The most commonly used varactor is a p–n junction diode which is specially doped to give large changes in capacitance between forward and reverse bias conditions. By applying a signal to the diode at a much lower frequency than the microwave frequency required, the variation in diode reactance introduces a large harmonic content in the output, and the desired microwave frequency can be selected by tuning. Varactors are highly efficient in producing microwave power at harmonics of a lower driving signal.

Of the newer methods of producing microwave power directly from a d.c. bias source, the Gunn oscillator (named after J. B. Gunn, who first demonstrated a practical circuit working on the principles to be described for this device), and the silicon avalanche diode or impatt oscillator (impatt being coined from *imp*act *a*valanche *t*ransit *t*ime) are the most promising and these will be briefly described.

The Gunn Diode

The Gunn diode consists simply of a single crystal of suitable semiconducting material sandwiched between two ohmic contacts, as shown in Fig. 11.6a. It is a diode in the sense of having two electrodes, but it does not have rectifying junctions of any sort. When the *direct voltage*

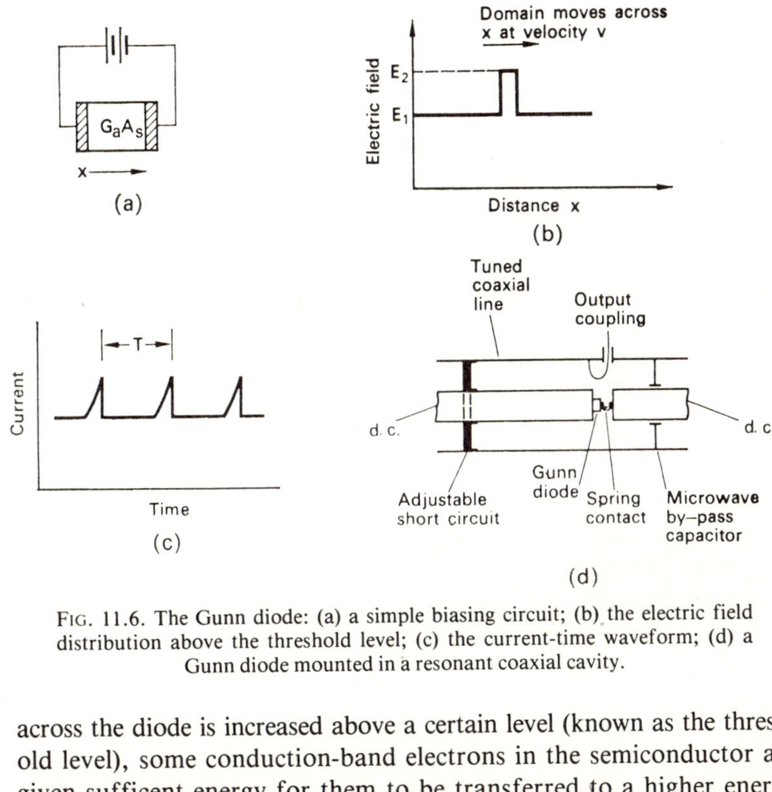

FIG. 11.6. The Gunn diode: (a) a simple biasing circuit; (b) the electric field distribution above the threshold level; (c) the current-time waveform; (d) a Gunn diode mounted in a resonant coaxial cavity.

across the diode is increased above a certain level (known as the threshold level), some conduction-band electrons in the semiconductor are given sufficent energy for them to be transferred to a higher energy level, into what is termed a sub-conduction band (see § 1.2 for description of the simple conduction band). Thus, for use as a Gunn oscillator, a semiconducting material requires the rather unusual energy band structure in which a second conduction band is available.

The effective mass of an electron (see § 1.8) in the higher energy band is considerably greater than that in the normal conduction band with the result that it has a lower mobility. For example, in gallium arsenide (GaAs) at 290°K, the figures are: normal conduction band, mobility 9000 cm^2/V sec, and effective mass 0.07 m_0, where m_0 is the rest mass; sub-band, mobility 150 cm^2/V sec, and effective mass 0.4 m_0.

Since all the conduction electrons must drift through the diode at the same velocity, those with low mobility bunch together to form an electric field domain. Representing the field strength across the domain by E_2, the drift velocity is given by $v = \mu_2 E_2$, where μ_2 is the lower value of mobility. Likewise, the higher mobility electrons must drift in a field of strength E_1 to satisfy the equation $v = \mu_1 E_1$. The total electric field distribution is sketched in Fig. 11.6b. While the electric field domain is present in the material, the current as seen in the external circuit, drops from its threshold value to some lower value, as shown in Fig. 11.6c. the periodic time T, of the current pulse (Fig. 11.6c), is given approximately by $T = L/v$, where L is the length of the diode, and v is the drift velocity. For GaAs, v is found to be of the order of 10^7 cm/sec, and for a typical diode length of 50 μm (or 5×10^{-3} cm), T is 5×10^{-10} sec. Thus the repetitive frequency of the current pulse, given by $f = 1/T$, is 2 GHz.

To increase the frequency of oscillation of the diode, the length L would have to be decreased, and this leads to difficulties in mechanical construction. Usually, an epitaxial layer is used to obtain very thin diode lengths, and the practical problem then becomes one of removing heat from this layer through the substrate. In a modified mode of operation, known as the limited space-charge accumulation mode (or LSA mode) the electric field domain is not given sufficient time to break up into the regions, as shown in Fig. 11.6b; in effect, the diode length is shorter than the domain width. The frequency of oscillation is then determined by the build-up time and quenching time of the domain, and is relatively independent of the diode length. The LSA-mode device can be operated at much higher powers than the normal Gunn diode, and is more efficient.

In practice, the diode is mounted in a resonant circuit, Fig. 11.6d (for either type discussed), which is tuned to the desired frequency, usually the fundamental component of the waveform shown in Fig. 11.6c.

The Silicon Avalanche Diode

Basically, the silicon avalanche diode is a *p–n* junction diode reverse-biased into the avalanche region. The voltage–current characteristics for a diode are shown in Fig. 11.7, and the avalanche region is where breakdown occurs due to impact ionization, and the current suddenly rises as shown at *A*, Fig. 11.7. The operation of the oscillator will be described with reference to a p^+-n junction (p^+ means very heavily doped with a *p*-type impurity), although there are other doping profiles that can be used and that offer certain advantages.

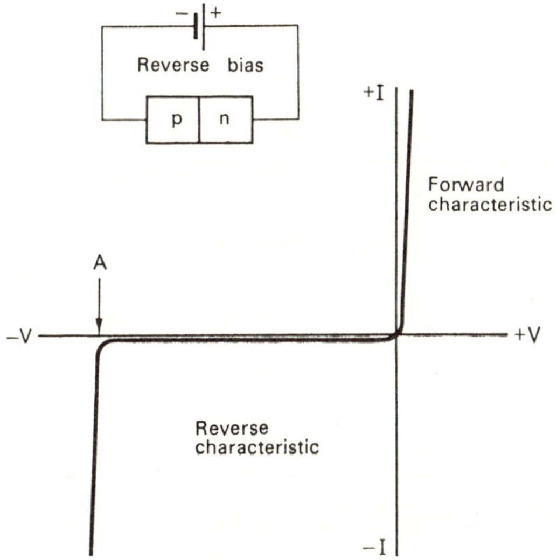

FIG. 11.7. The *V/I* characteristics of a *p–n* junction showing the avalanche breakdown region *A*.

When a p^+-n diode is reverse-biased (Fig. 11.8a), the depletion region (see § 1.5) extends almost entirely into the *n*-type material, and the electric field profile is as shown in Fig. 11.8b. By setting the d.c. bias near the avalanche threshold, and superimposing on this an alternating voltage, the diode will swing into avalanche conditions during alternate half-cycles. The hole–electron pairs generated as a result of avalanche action make up the current, with the holes moving into the

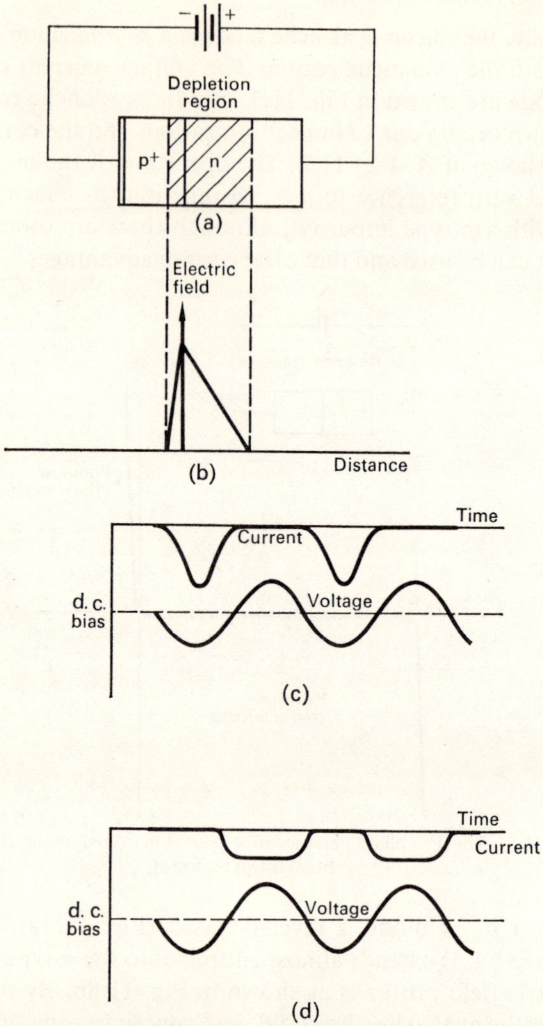

FIG. 11.8. The p^+-n silicon avalanche oscillator: (a) a simple biasing circuit; (b) the corresponding electric field distribution; (c) the current and voltage waveforms at low frequencies, when biased near the avalanche region; (d) the current and voltage waveforms at high frequencies.

p^+-region, and the electrons into the n-region. Because the electrons have a comparatively larger distance to travel through the depletion region they may still be present and drifting in the depletion region when the alternating voltage reverses. (It should be realized that under all conditions the diode is reverse-biased, the alternating voltage swinging about the d.c. bias.) At low frequencies, where the build-up time

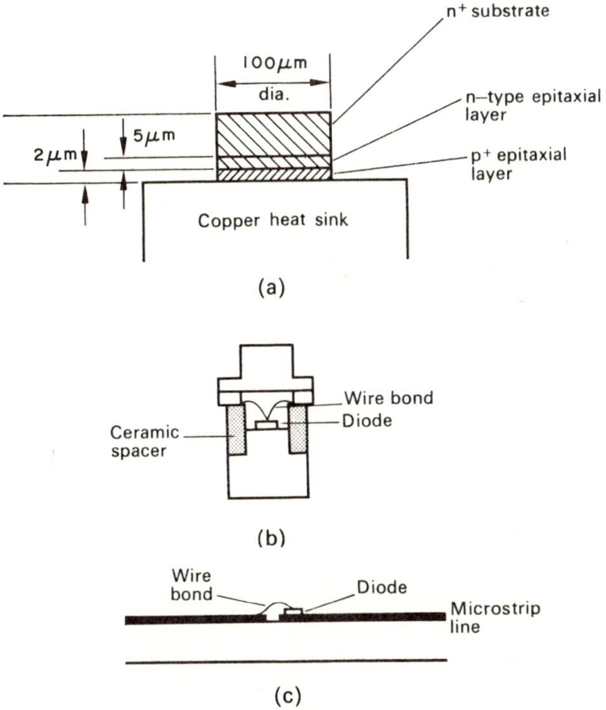

FIG. 11.9. (a) Typical dimensions for a silicon avalanche diode; (b) the diode in a coaxial mount; (c) the diode mounted on a microstrip line.

for the current, and the drift time through the depletion region together are negligible compared with the periodic time for the alternating voltage, the current appears simply as a "rectified" half-pulse (Fig. 11.8c). At high frequencies, where the total time lag for the current is comparable with the periodic time of the voltage, the current pulse will lag on the voltage. By making the drift time of the electrons in the depletion region equal to one-half the periodic time of the voltage, the current is in antiphase, as shown in Fig. 11.8d. This shift in phase of the current with respect to the voltage represents a negative a.c. resistance, so that the diode will sustain oscillations when placed in a resonant circuit (and the oscillations will build up from spontaneous fluctuations in carrier movement).

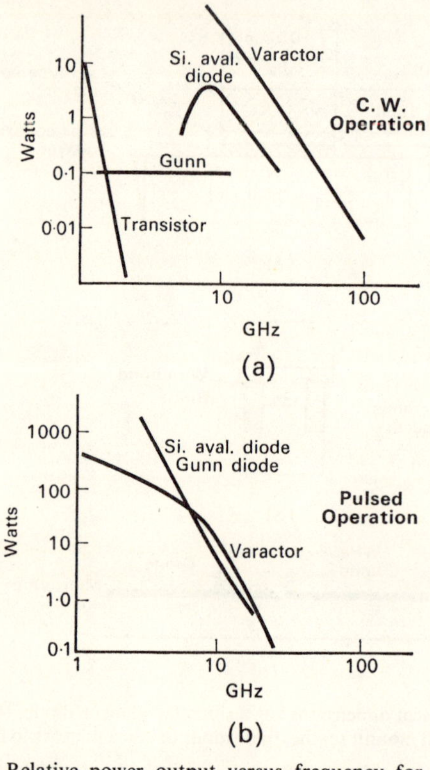

FIG. 11.10. Relative power output versus frequency for various types of solid-state microwave oscillators: (a) C.W. operation; (b) pulsed operation.

Figure 11.9a shows the dimensional details of a silicon avalanche diode, and it will be seen that the junction can be mounted directly onto a heat sink, thus facilitating heat removal (power dissipation can be of the order of 10^5 W/cm^2). Figure 11.9b shows how the diode may be mounted in a coaxial mount, and Fig. 11.9c shows how it may be connected into a microstrip line.

In Fig. 11.10, the relative power outputs for various types of microwave power generators are summarized in graphical form.

11.4. High Frequency Transistors

In order to get useful output power at high frequencies (typical state of the art values are about 80 watts at 30 MHz to about 5 watts at 4 GHz) transistors must be designed to have a higher periphery-to-area ratio for the emitter than that obtainable using a simple stripe geometry (as discussed in § 3.2). Essentially, area must be reduced without reducing periphery, as large area means large inter-electrode capacitance and hence decoupling at high frequencies.

With bipolar transistors the two most popular approaches to the problem are the interdigitated structure, sketched in Fig. 11.11a, and the overlay structure, Fig. 11.11b. In the interdigitated structure, a number of emitter stripes, connected in parallel, are interleaved with similar base stripes. The emitter aspect ratio *a* (see § 3.2) is equivalent to that of one long emitter the total length of which is the sum of the individual stripe lengths, and the width of which is that of a single stripe. One single long emitter stripe would be impracticable largely because its resistance would introduce sufficient voltage drop to prevent the emitter-base junction from being forward-biased towards the tip of the stripe. Ballast resistors are included in series with the emitter stripes to maintain a reasonable uniform current distribution. Otherwise two effects can occur which may damage the transistor. Heat flow is such that the temperature of the centre section of the transistor is higher than that at the edges, and as a result, current is also greatest through the centre section, which can have a cumulative effect on the temperature-current dependence. Secondly, channeling of current in the transistor can result in "second breakdown" in which high current density spots cause burnout in the base region ("first breakdown" is where avalanching occurs in the collector-base junction and is not normally a problem). By limiting the current to the various emitter sections the ballast resistors provide greatly improved protection from these effects.

The overlay structure is similar to the interdigitated structure in that the emitter consists of a number of stripes (or in some designs, emitter squares are used). However, these are connected in parallel by means of a metallization layer which "overlays" the base, as shown in Fig.

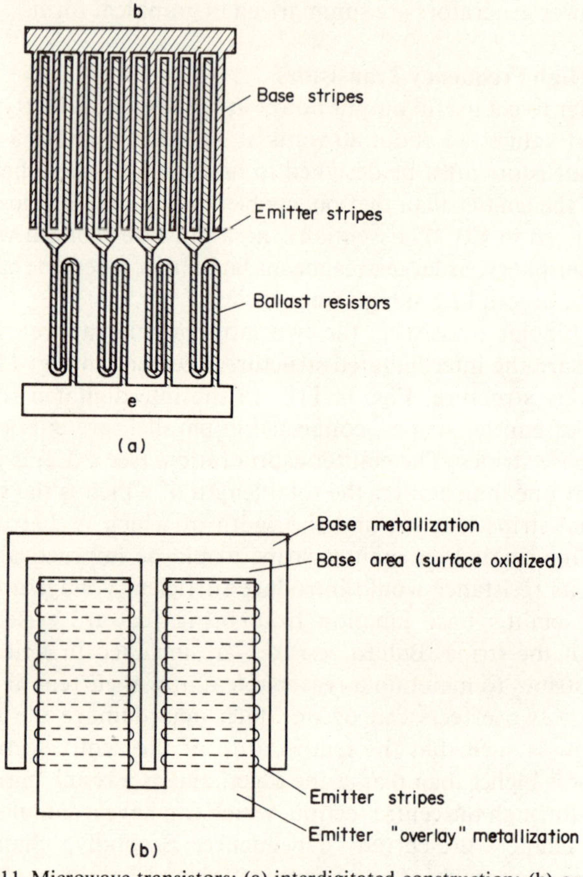

FIG. 11.11. Microwave transistors: (a) interdigitated construction; (b) overlay construction.

11.11b, there being an oxide layer to provide insulation between overlay and base. Commercial types are available with emitters ranging from as few as 16, to as many as 312. To date, both interdigitated and overlay types are manufactured in silicon, and overall dimensions are of the order of a few millimetres square.

Microwave field effect transistors are discussed in the next section.

11.5. Metal–Semiconductor Junction Devices

11.5.1. THE SCHOTTKY BARRIER DIODE

The Schottky barrier diode (sometimes termed the hot carrier diode) relies for its action on the rectifying properties of some types of metal–semiconductor junctions, as described in § 1.6, and it possesses a number of advantages which makes it highly suitable for microwave applications. It is a majority carrier device and therefore does not suffer from stored carrier effects (see § 3.2). It can be made having very

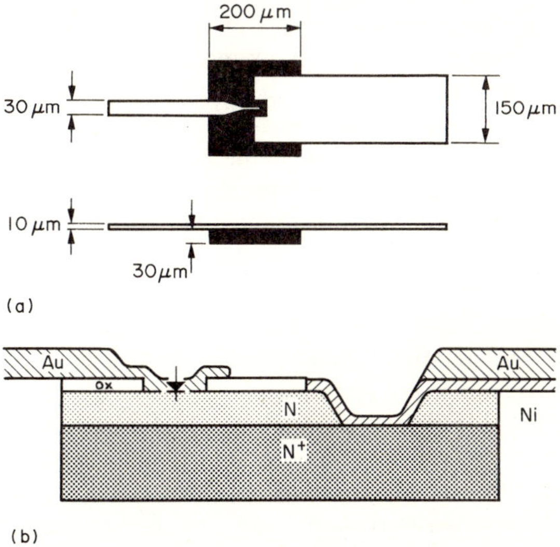

FIG. 11.12. Schottky barrier diode with beam-lead connections: (a) plan and side view; (b) enlarged cross-section. (Courtesy: *Philips Technical Review,* J. H. C. Van Heuven and A. G. Van Nie.)

small junction areas, and hence low capacitance. Further advantages are that it is easily fabricated in monolithic form, it only being necessary to deposit (usually by vacuum evaporation) the electrode onto an epitaxial layer of semiconductor. The capacitance-voltage characteristics of the junction provide good varactor action (variable reactance), and the characteristic is easily controlled by tailoring the shape of the electrode.

Figure 11.12 shows one form a Microwave Schottky diode may take, and also illustrates the beam-lead method of mounting (see also Fig. 10.5), which is suitable for connecting low-power devices into microstrip lines. The main dimensions of the diode are shown in Fig. 11.12a, and an enlarged cross section is shown in Fig. 11.12b.

11.5.2. GALLIUM ARSENIDE FIELD EFFECT TRANSISTORS

Gallium Arsenide (GaAs) is a particularly suitable material for high frequency devices. Both electron mobility and energy band gap are higher than the corresponding values in silicon (see Table 1.1). High mobility makes for high frequency or fast switching devices (bearing in mind however, that surface mobility will be considerably lower than bulk mobility), and high energy band-gap means that an intrinsic substrate will be semi-insulating; this can be doped with chromium which creates trapping centres and improves the insulating properties. Progress in GaAs has been slower than in Si mainly because it is a more difficult material to work with. Now, however, it is possible to grow epitaxial layers, doped to give a required conductivity, on a GaAs substrate, and also to form such layers by the alternative method of ion implantation doping. Both methods have been used successfully in producing GaAs integrated circuits.

The basic GaAs FET is shown in Fig. 11.13a. Ohmic source and drain contacts are formed (one method being to alloy in gold-germanium contacts). The gate metal is deposited directly on the *n*-type layer to form a Schottky barrier contact (see § 1.6). By increasing the magnitude of the negative bias between gate and channel (connected through the source contact), the reverse-bias depletion of the Schottky diode constricts the conducting channel between source and drain. Output characteristics are sketched in Fig. 11.13b. The circuit symbol for the GaAs FET is shown in Fig. 11.13c, which shows that the input gate-channel is a diode (compare with the IGFET symbols shown in Fig. 6.7). The arrow as shown in Fig. 11.13c indicates an *n*-channel device, it indicating the direction in which conventional current would flow if the diode were forward-biased. The GaAs FET comes into the class of devices known as junction gate field effect transistors or JGFETs.

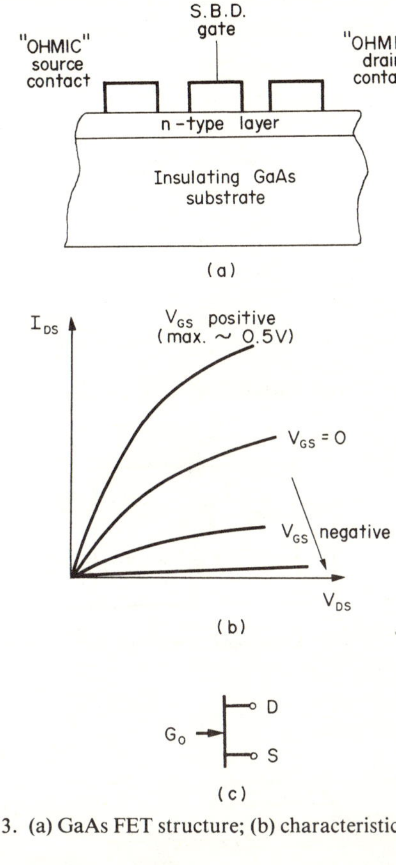

FIG. 11.13. (a) GaAs FET structure; (b) characteristics; (c) symbol.

Being a majority carrier device, charge storage effects (see § 3.2) are absent. Also, the comparatively simple geometry eases the problem of achieving narrow channel widths, and widths of the order of 1–2 μm are possible in practice. No overlap occurs between gate and source and drain electrodes, so that interelectrode capacity is minimized.

Microwave amplifiers in both hybrid, and monolithic forms, have been constructed (see for example S. J. Hewitt and R. S. Pengelly, Design Techniques for Integrated Microwave Amplifiers using GaAs FETs, *The Radio & Electronic Engineer,* October 1976). In the hybrid circuit, the transistor is mounted in a l.i.d. (leadless inverted device) package as shown in Fig. 11.14a. The equivalent circuit for the assembly

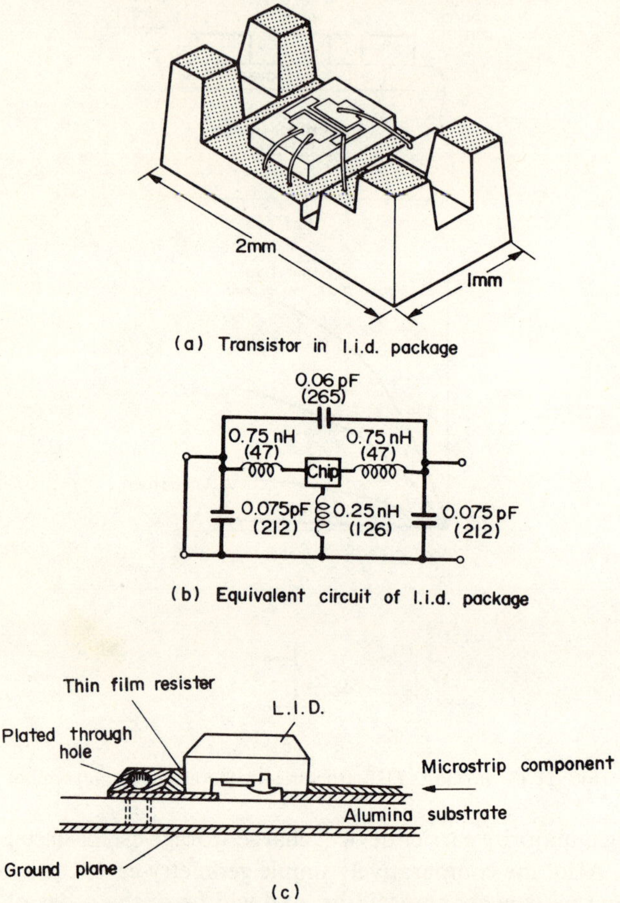

FIG. 11.14. (a) Transistor in l.i.d. package; (b) equivalent circuit of l.i.d. package; (c) mounting of l.i.d. package, and connection to ground plane by means of plated-through hole. ((a) and (b) Courtesy: *The Radio & Electronic Engineer*, S. J. Hewitt and R. S. Pengelly.)

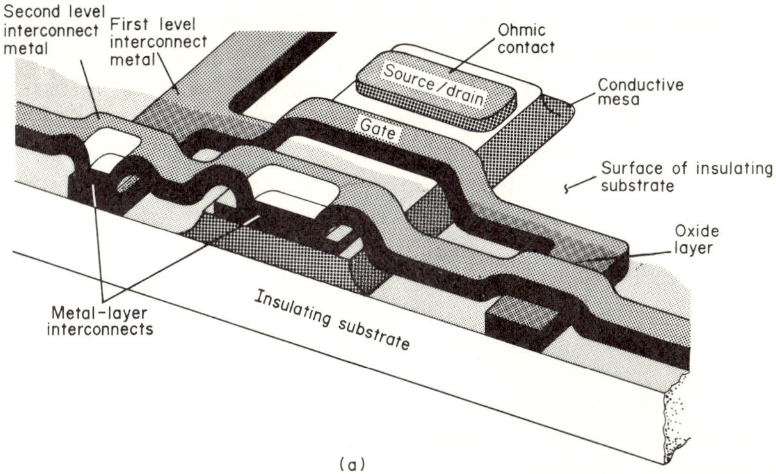

(a)

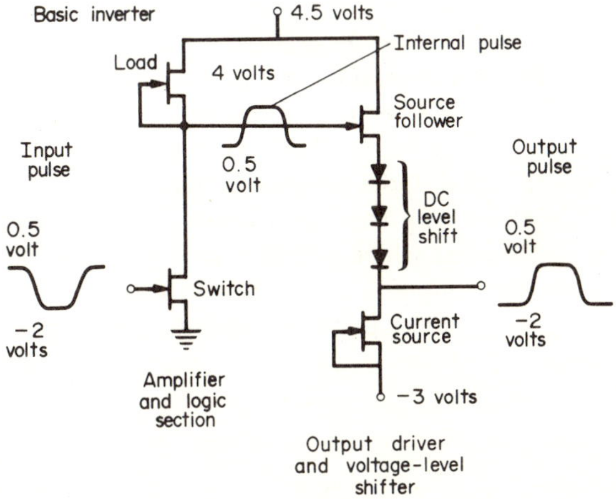

(b)

FIG. 11.15. GaAs logic circuits: (a) constructional details; (b) basic inverter circuit. (Courtesy: *IEEE Spectrum*, Rory Van Tuyl and Charles Liechti.)

is shown in Fig. 11.14b. The package is connected into the circuit in the manner illustrated in Fig. 11.14c, this being similar to the "flip-chip" technique discussed in Chapter 10, except that in the l.i.d. package, bonding wires are required between device and the pads on the ceramic carrier. Connection to the ground plane is made by means of plated through holes as shown in Fig. 11.14c.

Logic switching circuits have also been fabricated on a single chip (see for example R. van Tuyl and C. Liechti, Gallium Arsenide spawns speed, *IEEE Spectrum*, March 1977). In this approach, mesas are formed for the active devices by etching away the *n*-type conductive layer as shown in Fig. 11.15a. The first layer of metal interconnect forms a gate wherever it crosses a mesa. An insulating oxide is deposited on top of the first metallization, and a second metal interconnect formed on top of the oxide. Holes are etched in the oxide where the second interconnect layer has to connect with the first (this is rather similar to the plate through hole technique shown in Fig. 11.14c except that here the second metal will deposit directly through the hole in the oxide). Through connections of this type are known as vias.

The circuit for a high speed logic fabricated in this manner is shown in Fig. 11.15b. The quoted time-delay product for the circuit is 100 ps × 20 mW = 2 pJ. The transistors operate in the depletion mode and as a result, d.c. level shifting is required to maintain compatibility between input and output voltage levels. The required d.c. voltage offset is obtained across the three series connected Schottky diodes shown in Fig. 11.15b. In the article by Tuyl and Liechti, a microwave divide-by-eight counter is described which operates with an input frequency of 2 GHz. The complete counter is fabricated on a gallium arsenide chip 800 μm × 880 μm, and is an example of medium scale integration (MSI) in this technology.

CHAPTER 12

Semiconductor Memories

12.1. Introduction

A system is said to have a memory whenever its response to a given input depends on the previous state of the system. For example, the ability of an electric capacitor to store charge can be used as a "memory" record of a voltage although in this case the memory gradually dies out and would have to be refreshed. Punched cards are an example of a comparatively permanent memory system, as is a book.

The commonly encountered memories in electronic systems (e.g. computers, microprocessors) are: punched cards and paper tape; magnetic tapes and discs; magnetic cores and bubbles; semiconductor circuit memories. Only the latter will be discussed in this chapter.

Memory systems are classed as *volatile*, and *non-volatile*, and most semiconductor memories come into the volatile class, that is, the information the memory stores is lost (or erased) if the power supply to the memory is disconnected. However, certain types of MOS memories can now store information in the form of trapped charge, which remains trapped even when power supplies are turned off. These types are discussed in § 12.5, and are an example of non-volatile memories. The other, more obvious non-volatile types of memories are the punched cards, or paper tape, and magnetic types, which retain the stored information in the absence of external power supplies.

12.2. Some Commonly Used Terms

A memory *cell* is the basic circuitry required to store one bit of information.

A *bit* is a binary digit i.e., either a logic $\underline{1}$ or $\underline{0}$.

A *word* is a group of bits which can be treated as a single unit. Each word occupies a complete memory location, for example, an eight-bit word would require eight separate memory cells, and the eight cells would be termed a memory location.

A *byte* is a *consecutive* sequence of bits which may be treated as a single unit; a *byte* may differ in bit length from a word in a system.

The *address* of a unit of information (bit or word) is its location in the memory. In the systems discussed here, the address is coded in the form of binary signals.

Access time is the time that elapses between applying an address signal at the address input and the information appearing at the memory output.

Memory *capacity* is the number of bits the memory can store. Although this will be some power of two in most binary systems, for capacities in the thousands of bits, it is common practice to refer to the capacity in terms of the thousands units only. For example, a 1024 (2^{10}) bit memory is referred to as a 1k memory; an 8192 (2^{13}) bit memory as an 8k.

Memory device refers to a complete memory system available as a single component in a package suitable for circuit board connection. The device chip within the package contains a large scale integrated (LSI) circuit which, in addition to the memory cells includes circuitry for controls, decoders, clocks, sense and drive amplifiers. A variety of LSI packages are available, a typical example being the dual in-line package (DIP) illustrated in Fig. 12.8.

12.3. Read Only Memories (ROMs)

A Read Only Memory (ROM) is a circuit which has a number of *word line* inputs, and a binary output, and the binary output generated for a given word input follows rules determined by the wiring of the circuit. With no input, the word lines are at logic $\underline{0}$; a binary input is first passed through a *decoder* which results in one only of the word lines being raised to logic level $\underline{1}$. This word line causes the ROM (also known as the *encoder*) to respond with the required output. The ROM stores instructions only, it does not store data.

As a simple illustrative example, suppose a ROM is required to provide an output which is the square of the input, for the integers 3, 4, or 5. One possible arrangement is shown in Fig. 12.1. A binary input is converted to the corresponding word line signal by means of the inverters and AND gates, the correspondence between these being shown

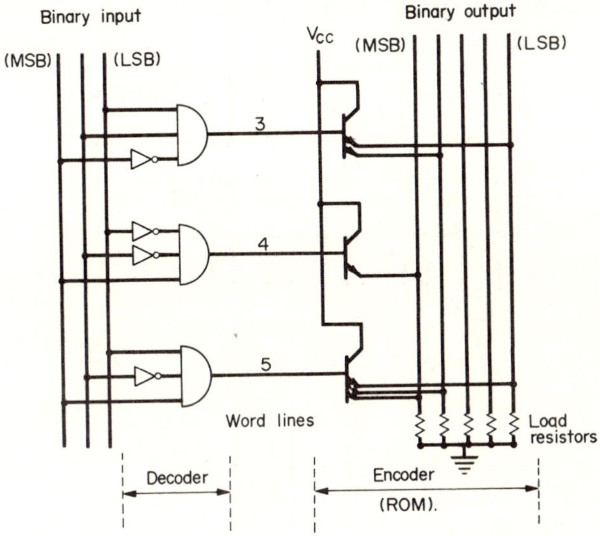

Input		Word lines	Output	
Decimal EQ.	Binary (MSB) (LSB)	5 4 3	Binary (MSB) (LSB)	Decimal EQ.
3	0 1 1	0 0 1	0 1 0 0 1	9
4	1 0 0	0 1 0	1 0 0 0 0	16
5	1 0 1	1 0 0	1 1 0 0 1	25

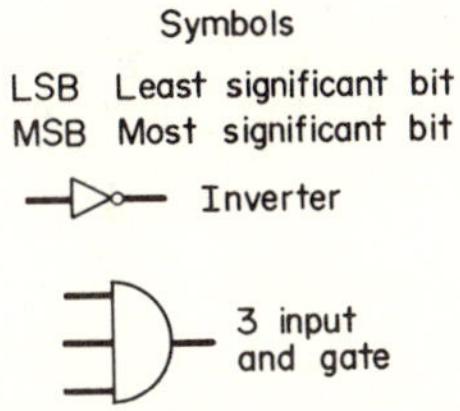

FIG. 12.1. A simple ROM function.

in the truth table of Fig. 12.1. The word line at logic level 1 switches on the multi-emitter transistor to which it is connected, and the emitters, which are connected in emitter-follower configuration to the load resistors generate logic level 1, the other output lines remaining at logic level 0. For example, word line 3 turns on the two-emitter transistor which in turn puts a logic level 1 on the 2^3 and 2^0 output lines, the other output lines remaining at logic level 0.

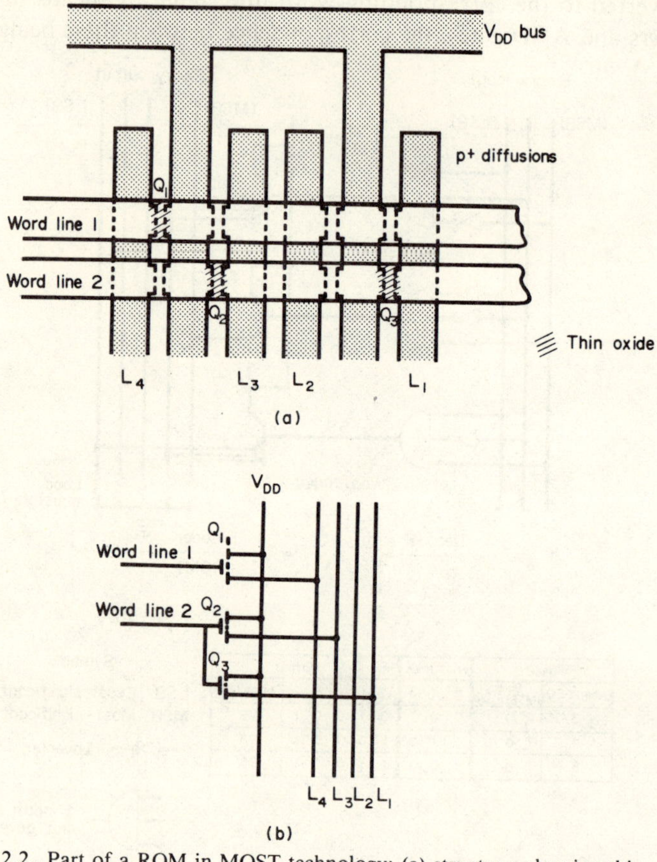

FIG. 12.2. Part of a ROM in MOST technology: (a) structure, showing thin oxides for gates, and thick oxides for crossovers; (b) equivalent circuit for (a).

MOST technology is particularly suited to making large ROM arrays. Figure 12.2 illustrates a commonly used method. A thick oxide (~10,000 Å) comes between the word lines and the p^+ diffusions except where MOSTs are required. Here the thick oxide is etched away and a thin ($\sim$1000 Å) gate oxide deposited, as shown for Q_1, Q_2, and Q_3, Fig. 12.2a. The circuit for this particular layout is shown in Fig. 12.2b. The thick oxide is also used to isolate metallization layers where these cross-over and a connection is not desired. In this case, a p^+ diffusion is used to "tunnel" under one of the metallization layers, and the connection for the second layer is made via the tunnel.

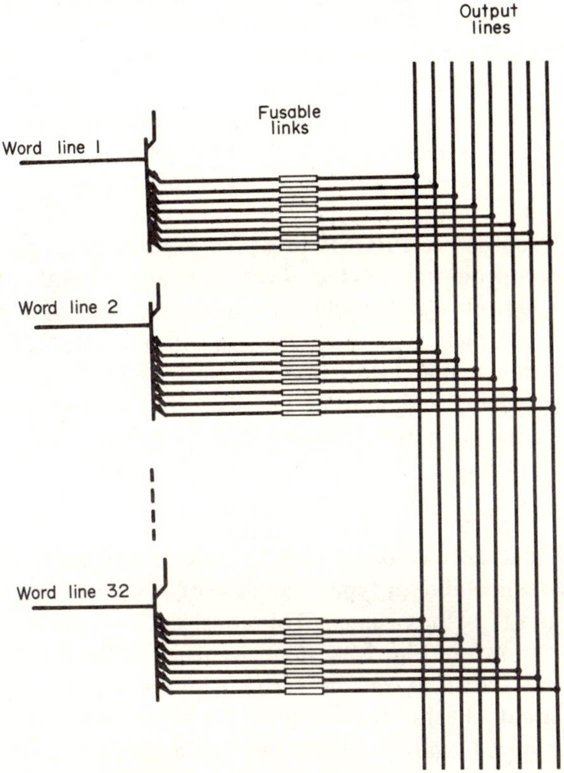

FIG. 12.3. Programmable ROM.

As can be seen, the particular function for a ROM is "hardwired" in. They are said to be *mask programmed* because the masks used in manufacture (see Chapter 2) are designed specifically for the function. Where demand is sufficient to justify mass-production, as for example with calculator chips, ROMs can be bought already programmed. For small production runs and prototypes, such as are required for some microprocessor applications, programmable ROMs (known as PROMs) are available. These are also known as field programmable ROMs, meaning that they can be programmed by the user in his particular application. PROMs are bipolar arrays in which all the emitter encoder connections are fused as shown in Fig. 12.3. The user determines which connections are required for his program, and open-circuits the unwanted connections by fusing the links ("zapping", in the jargon). Figure 12.3 shows part of the encoder section from a commercially available PROM which has 32 eight-emitter transistors. The decoder accepts a 5-bit input code which it converts to a 32 word-line output, which in turn is encoded onto the eight bit output by means of the user's program.

Erasable PROMs are PROMs in which existing programs can be erased and new programs entered. Two types are presently available, one of which uses ultra-violet light for erasing and entering a program, this type being known as an EPROM (for erasable PROM); and the second type, known as an EAPROM, for electrically alterable PROM, in which the erasing and writing are achieved by electrical means. These are described in more detail in § 12.5.

12.4. Random Access Memories (RAMs)

The random access memory is used for storing information in digital form. Many semiconductor types make use of a basic memory cell circuit known as a flip-flop or *latch*. This consists of two inverters cross-connected as shown in Fig. 12.4. The circuit is stable in either of the two conditions shown as can be verified by tracing the logic levels around the circuit. In the RAM, one of the states shown represents a logic 1 being stored, and the other state represents a logic 0, and additional circuitry must be provided for changing the state as required, and for determining or reading, the state the flip-flop is in. Typical circuits will be described shortly, but the point being made here is that the flip-flop remains in whichever state it is switched to providing the power supplies (but not necessarily the switching signals) are kept on. A *truth table* is a tabulation of the logic states, and this is shown for the flip-flop in Fig. 12.4c. Note that it is not necessary to show the $\bar{A}$ and

$\bar{Q}$ levels as these are the complements of A and Q respectively (if $A = 1$ then $\bar{A} = 0$ etc.).

Figure 12.5a shows how arrays of flip-flop cells are arranged for what is termed *coincident addressing*. Each board shown in Fig. 12.5a contains 16 cells, as shown in more detail in Fig. 12.5b. The X and Y address decoders generate X and Y word lines for a given X, Y binary

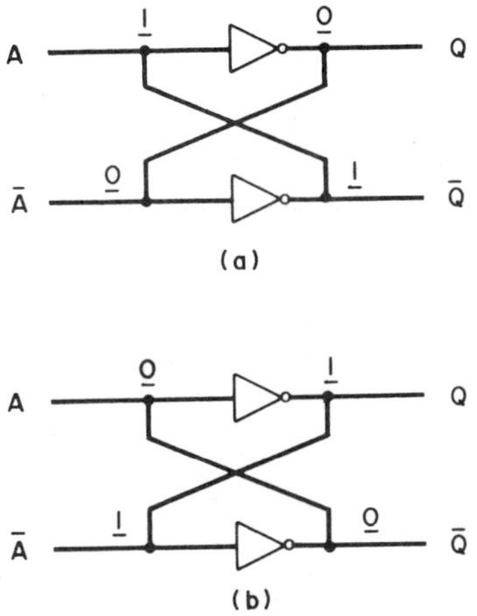

A	Q
1	0
0	1

(c)

FIG. 12.4. The basic flip-flop, (a) and (b) showing the two stable states; (c) the truth table.

inputs in the same manner as for the ROM, so that only one X word and one Y word line are at logic level 1 for any address. The non-addressed lines remain at logic level 0. (For the circuits to be described, logic level 1 will be at or near V_{cc} or V_{dd}, and logic level 0, at or near ground potential.) The cell at which the addressed X and Y word lines coincide is the only cell activated. The write lines W and $\bar{W}$ are used to change the state of the cell, i.e. to write in data, and the sense lines S and $\bar{S}$, to determine the state the cell is in.

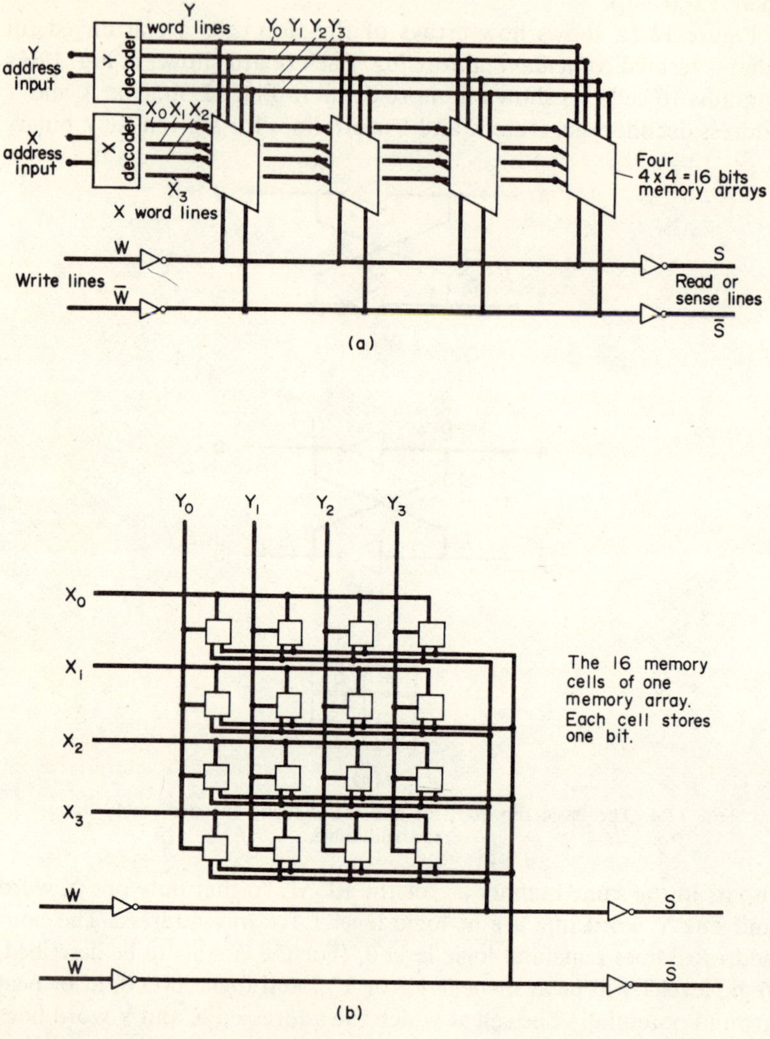

FIG. 12.5. (a) Overall memory organization; (b) coincident addressing of memory cells, for a RAM.

Figure 12.6a shows the commonly used bipolar transistor type cell. This has two 3-emitter transistors, and two of the emitters for each transistor are lifted to logic level 1 by the address lines. To write in a logic 1 the write lines are set at $W = 1$ and $\overline{W} = 0$. The two driver inverters

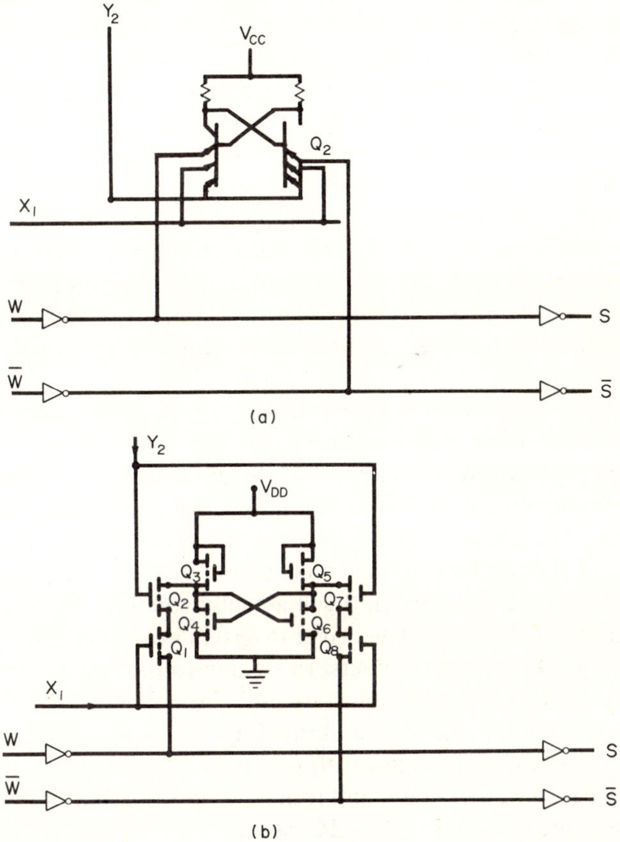

FIG. 12.6. Static logic cells for a RAM: (a) bipolar type; (b) MOST type.

in the write lines invert these so that the remaining emitter at Q_1 is biased 0, while that at Q_2 is biased 1. Thus Q_1 conducts Q_2 is turned off, and the resulting rise in Q_2 collector voltage maintains a forward bias on the base of Q_1 which keeps this on even when the address and write signals are removed (remembering that removal of these returns the lines to $\underline{0}$ level).

To write a $\underline{0}$ into the cell, it is addressed as before, and $W = \underline{0}$, $\overline{W} = \underline{1}$ applied. Now Q_2 is made to conduct and Q_1 is turned off with the result that the circuit moves into its alternate stable state. It is seen therefore, that when a $\underline{1}$ is being stored, Q_1 is conducting and Q_2 is off, while a $\underline{0}$ being stored means that Q_1 is off and Q_2 conducting.

To read the state of the cell, it is addressed as before. The off transistor will be unaffected by the address signals (since it is already off) while in the on transistor, the emitter currents to the address lines are now diverted into the remaining emitter at level $\underline{0}$, which is connected to the sense inverter. This inverter will turn on, and hence its output drops to logic level $\underline{0}$. The output conditions that result are: when a $\underline{1}$ is stored, $S = \underline{0}$ and $\overline{S} = \underline{1}$; when a $\underline{0}$ is stored, $S = \underline{1}$ and $\overline{S} = \underline{0}$. Note that read-out is achieved without destroying the information being stored.

The basic MOST cell is shown in Fig. 12.6b. This consists of two NAND–NOR gates connected in a flip-flop arrangement. Although there are eight transistors in the cell, the diffusions for many of these are common, similar to that shown in Fig. 7.6 and the cell area is not excessive. The address signals bias the NAND branches, Q_1,Q_2 and Q_7, Q_8, of the two gates to the on condition, but only the branch which is connected to a write line at $\underline{0}$ will actually conduct. To write in a $\underline{1}$, W is set at $\underline{1}$ and $\overline{W}$ at $\underline{0}$. The driver inverters invert these as before so that the Q_1,Q_2 branch conducts, this going to the line at level $\underline{0}$. The left-hand gate, being in the on state pulls the bias on Q_6 in the right-hand gate to below cut-off; the branch Q_7,Q_8 is non-conducting because it is returned through Q_8 to the line at level $\underline{1}$. Thus, the right-hand gate is in the off state, and it maintains an on-bias on Q_4 of the left-hand gate even when the address and write signals are removed.

To write in a zero, W is set at $\underline{0}$ and $\overline{W}$ at $\underline{1}$, and the address signals applied as before. Following similar reasoning for the write $\underline{1}$, it will be seen that Q_4 is turned off while Q_6 is turned on, and this condition is stable even when the address and write signals are removed. Thus, a $\underline{1}$ is stored when Q_4 is on and Q_6 off, and a $\underline{0}$ when Q_6 is on and Q_4 off.

To read the state of the cell, it is addressed as before, and the gate which is on will pull the write/read line to which it is connected to the $\underline{0}$ level. For example, if a $\underline{1}$ is stored (Q_4 conducting) the level at the load transistor Q_3 is $\underline{0}$, and this is transferred through Q_1, Q_2 (switched on by the address signals) to the top line, while in a similar manner, the $\underline{1}$ level at load transistor Q_5 is connected through Q_7, Q_8 to the lower line. The sense amplifiers invert these so that for a $\underline{1}$ stored, $S = \underline{1}$ and $\bar{S} = \underline{0}$. For a $\underline{0}$ stored, $S = \underline{0}$ and $\bar{S} = \underline{1}$ when the cell is addressed. Again, note that readout does not destroy the stored data.

The memory cells of Fig. 12.6 employ what is termed *static logic.* By this is meant that once the data is written into the cell, it will remain there while power is on, without further attention. Another form of logic widely used is *dynamic logic.* With this, the logic state is determined by the state of charge of a capacitor in the cell. Dynamic logic requires fewer components per cell than static logic, but the state of capacitor must be "refreshed" periodically, typically once every 2 ms, which requires more complex peripheral circuitry. However, large RAM memory arrays use dynamic logic (e.g. 16k arrays) as the much smaller cell size permits larger cell density per chip. In recent designs, the storage elements and switching capacitors have been merged in one-cell structures.

The merged transistor cell is briefly described in § 4.7 (see also Fig. 4.12). This is used by the Fairchild Camera & Instrument Corp. in their model 93481 4k RAM. The memory is organized in 32 by 128 bit words. For each of the 32 words a pair of lines is needed, W_p and W_n of Fig. 12.7a. The address input is a 7-bit word, the first five bits of which are applied to a word decoder to select a pair of word lines ($2^5 = 32$), and the full seven bits to a bit decoder to select the desired bit line ($2^7 = 128$).

Bit data are stored in the base-collector capacitances. The uncharged state represents a stored logic $\underline{1}$, the charged state (with a bit line positive) a logic $\underline{0}$. Apart from the refresh logic $\underline{1}$ period to be described shortly, the W_p line is kept near ground potential which reverse biases the emitter-base junction of the *p–n–p* transistor.

Data is written into the cell by first selecting the cell through the word and bit line addresses. W_n is reduced to near zero potential (the same as W_p) during the write operation. To write in a logic $\underline{0}$, the bit line is raised to a high level, (about 3.4 V) by means of a latch (flip-flop) sense amplifier. A charging current path now exists from the bit line,

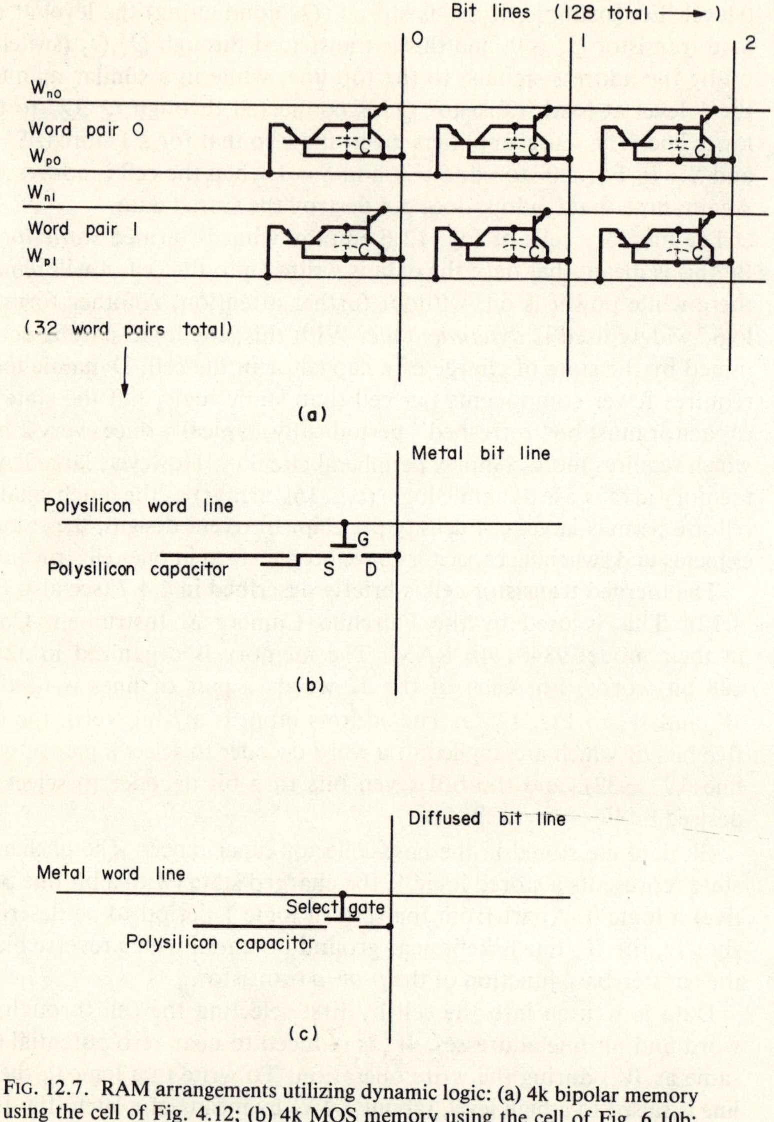

FIG. 12.7. RAM arrangements utilizing dynamic logic: (a) 4k bipolar memory using the cell of Fig. 4.12; (b) 4k MOS memory using the cell of Fig. 6.10b; (c) 16k MOS using the cell of Fig. 6.10c.

through C and the base-emitter junction of the n–p–n transistor to the low W_n line, and capacitor C is charged.

To write in a logic 1, the bit line is reduced to near ground potential by means of the sense amplifier latch. Both terminals of C are now at the same low potential and C discharges to zero.

Data is read out of the cell by means of the sense amplifier latch. As for the write operation, both W_n and W_p are near ground potential, and the bit line is high (about 3.4 V). If C is uncharged (logic level 1) the n–p–n transistor will conduct. Capacitance C is multiplied by the transistor β (Miller effect), with the result that the W_n line, at low potential, is capacitively coupled to the bit line, which reduces the bit line potential sufficiently for the sense flip-flop to change state, i.e. the bit line is pulled low.

If C is charged (logic level 0), the voltage across C prevents the emitter-base junction of the n–p–n transistor from being forward biased. The capacitance coupling is therefore reduced by a factor of about β, which is insufficient to alter the bit line potential, which is held high by the sense latch. Typically, C is about 0.1 pF, and β about 70, which enables logic swings of 200 mV to be detected.

In practice, C will have discharged somewhat as a result of leakage, and a small charging current will flow during the read 0 operation, sufficient to refresh C, without altering the latch setting. Thus, logic level 0 is automatically refreshed during the read cycle.

The refresh 1 operation is carried out at the end of each timing cycle. When reading a logic level 1, the sense amplifier remains latched at its low level for a brief period after W_n has returned to the normally high level (3.4 V). W_p is raised to the high level during this brief period. Any stray charge on C (which would make the bit line positive) now provides a collector current source for the p–n–p transistor which is turned on by W_p, and discharges C.

Of course the read/refresh operations must be carried out on a periodic basis, typically, each line must be refreshed once every 2 ms.

Dynamic logic memories in MOS technology are also available which utilize charge storage in a capacitor to represent the logic levels. Figure 12.7b shows the bit and word line arrangement for a 4k memory which utilizes the cell structure shown in Fig. 6.10b. Here, transfer of charge to and from the capacitor to the bit line is made through the

MOST. A logic swing of about 200 mV, comparable to the MTL cell of Fig. 12.7a, is achieved. An advantage claimed for the MTL structure over the MOS structure is that it is about twice as fast for very little increase in power consumption.

The MOS technique has been further developed to produce a 16k memory for which no counterpart in bipolar technology exists at the present (1977). The bit and word line connections to a cell are shown in Fig. 12.7c, and is seen to be similar to the 4k version. The cell structure, which is shown in Fig. 6.10c, differs markedly from the 4k cell. No source diffusion is used, a diffused bit line is used, and the select gate allows charge to be transferred to and from the bit line to under the capacitor line. The cell area for the 16k memory is about one half that of the 4k cell, being about 450 $(\mu m)^2$, but it still produces a logic swing of about 200 mV.

12.5. Non-volatile Semiconductor Memories (NVSMs)

Non-volatile semiconductor memories are devices which retain the data stored in them when external power supplies are switched off. The retention characteristic of all presently (1977) available types is achieved by alteration of the threshold voltage of IGFETs forming part of the memory. A low threshold voltage corresponds to the *erased* state, that is, the device is easily switched on and off by applied gate voltage. A high threshold voltage corresponds to the *programmed* state, in which the device cannot normally be switched on. The difference between the threshold voltages is termed the *threshold window*.

To be of practical use the retention time for a particular threshold state must be of the order of months or even years, as has been measured for some devices. The threshold window should be at least two volts and should be stable for repeated switching for at least 10^6 operations. Threshold switching time should be of the order of milliseconds or microseconds, and switching voltages are required to be less than 20 V to be compatible with existing integrated circuit technology.

Two types of threshold memory elements are in use, the MNOS (Metal–Nitride–Oxide–Silicon) transistor, and the floating-gate transistor. Considerable variation in design exists within each type, and both *n*-channel and *p*-channel units are available. A cross sectional

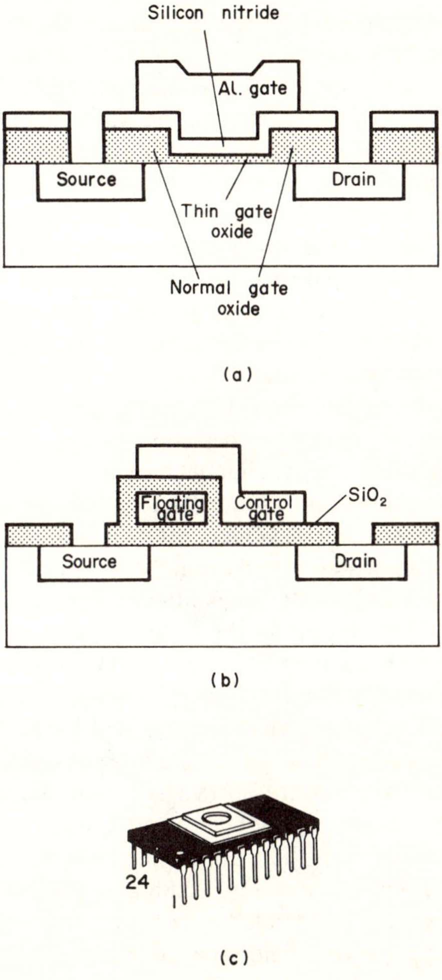

FIG. 12.8. (a) The metal-nitride-oxide-semiconductor transistor; (b) the floating gate transistor; (c) LSI package for ultra-violet light erasure of memory.

view of a MNOS transistor is shown in Fig. 12.8a. The SiO_2 layer is of the order of 30 Å thick, which allows carriers from the transistor channel to tunnel through to trapping states at the interface with the silicon nitride (Si_3N_4). The traps are distributed throughout the double insulator, but are mostly located at the interface. Tunnelling is a quantum-mechanical process, that is, the wave function representing carriers can spread into, and through, an insulator, a situation not possible in the discrete particle model of carriers.

To "write" into an *n*-channel device, the gate is made positive with respect to the channel. Electrons penetrate through to the interface and are trapped there, charging the oxide negatively. To "erase" the negative charge the channel is made positive with respect to the gate, and the electric field within the insulator results in electron emission from the traps back into the silicon.

It is possible to get hole tunnelling into the insulator which would charge this positively, resulting in depletion mode conditions. To avoid this, only the centre part of the channel is controlled by MNOS action, the channel regions adjacent to source and drain being normal enhancement-mode MOS structures. These MOS regions also ensure high breakdown voltages at source and drain.

The basic floating-gate MOS has already been described in § 6.5 (see Fig. 6.10a). The charge on the floating-gate performs the same function as the trapped charge in the MNOS device. Figure 12.8b shows the main features of a floating-gate transistor known as SIMOS (for Stacked-gate Injection MOS), developed by Siemens, Munich. The control gate is stacked on top of the floating-gate. Both gates are of polysilicon, and for erasure, ultra violet light will penetrate these and the insulator, to generate hole-electron pairs in the silicon. Some of the holes combine with the trapped electrons to neutralize these. The particular design shown in Fig. 12.8b also permits electrical erasure. The floating-gate is extended over the source region as shown, and an avalanche electric field applied across the source-channel junction. High energy holes (termed "hot holes") penetrate through to the floating-gate and neutralize trapped electrons. With this method it is possible to overcompensate for the negative charge (this does not happen with the UV method of erasure), and the floating-gate becomes positively charged, resulting in depletion-mode conditions. To ensure enhancement mode conditions at all times, the channel region adjacent to the drain is operated in normal enhancement MOS mode by the control gate as shown.

UV erasure usually entails erasing the complete memory chip, which may contain many thousands of memory cells. The package for a

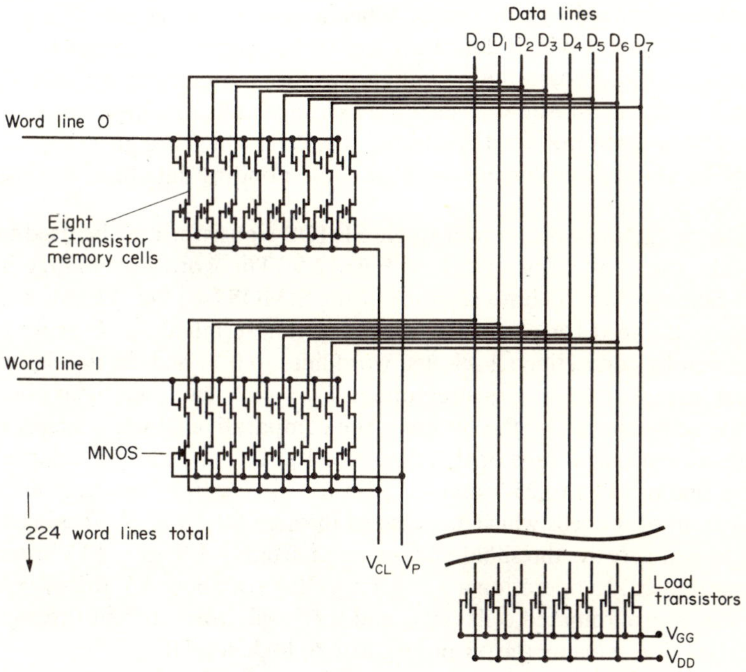

FIG. 12.9. Organization of a two transistor cell, electrically alterable read only memory, utilizing the MNOS transistor.

Motorola 8192-bit, UV erasable PROM, is shown in Fig. 12.8c, the window on the package being transparent to the UV radiation. Electrically alterable PROMs usually incorporate access to the cells on a word or bit basis so that words or bits can be erased without disturbing data to be retained.

Figure 12.9 shows in part the circuit of a 8192 bit EAROM organized in 1024 by 8 bit words. Each memory cell contains two transistors, one a normal enhancement mode MOS, the other a threshold alterable MOS. All cells go to the V_{CL}terminal and can be erased simultaneously through V_{CL}, by turning on all gates and all data lines so that a circuit through the cells is completed for V_{CL}. Alternatively, particular words can be selected through the appropriate word line. The device is programmed on word-by-word basis. When a word line is selected, the program determines which data lines are to be selected to complete the circuit for V_{CL}. Programming voltages are then applied to V_{CL} and V_P which alter the threshold levels in the selected cells to a high value. In operation, when the word is selected, the cells which are programmed "off" will not draw current, and the corresponding data lines will rise to logic level 1.

Figure 12.10a shows how a single SIMOS transistor may be used to replace the two-transistor cell of Fig. 12.9. The word-line voltage in this case operates on both sections of the SIMOS (see Fig. 12.8b), and there is no need for a separate programming terminal V_P. Erasure is achieved by connecting all selected word lines to 0 V, and the unselected word lines to +25 V. Erasure voltage $V_{CL} = 48$ V can then be applied to the cells selected by the bit lines (the comparatively high voltages is a possible disadvantage of this type of cell). To program, the selected word line is raised to 25 V and V_{CL} adjusted to give 16 V between source and drain of the cells which are selected through the bit lines. These cells are turned off, the threshold voltage rising from + 1 V to + 10 V when programmed. In operation a word line voltage of about 5 V will switch on the unprogrammed cells only, and these will draw current through the data line, causing the output to drop to logic level 0.

Figure 12.10b shows the cell circuit of a 256 bit memory cell (Nitron NCM 7040) which can be programmed in operation, and in this respect it is like a low-speed RAM. With the circuit operating as a flip-flop, a non-volatile data bit can be stored by programming one of the MNOS transistors, in the flip-flop load paths, to the off (actually low conductance) state, and the other to the high conductance state. In this way the "latch" current to the flip-flop is fixed.

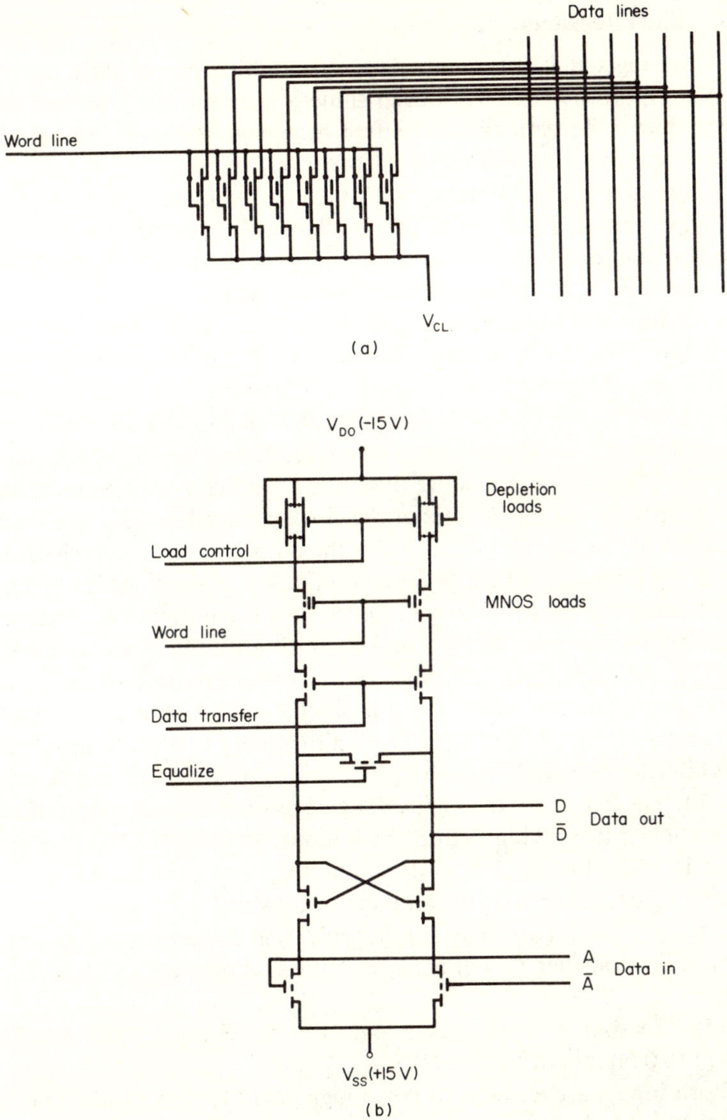

FIG. 12.10. (a) Part of the circuit for a single transistor cell, electrically alterable read only memory, utilizing the floating gate transistor; (b) use of the MNOS transistor in a flip-flop type memory cell.

12.6. Shift Registers

A shift register is a type of memory in which binary data can be stored temporarily while various arithmetic operations are carried out on the data. The data can be shifted along one storage cells within the register, and hence the name shift register.

As with RAMs, both static and dynamic logic may be used, and both bipolar and MOS types are available. In general, the bipolar types provide fastest operation while MOS types provide high density of storage cells at lower cost than the bipolar type.

Static logic types utilize the flip-flop, or latch, as the basic cell, but this has to be modified so that clocking signals can be applied to synchronize the shifting of data. Figure 12.11a shows the logic diagram for a clocked flip-flop. This is known as an S–R flip-flop, in which the S input is used to set the output Q to a logic $\underline{1}$, and the input R is used to set Q to a logic $\underline{0}$. The state of the flip-flop can only be changed when a clocking pulse is present at the clocking terminal CK. The truth table is shown in Fig. 12.11b, in which the output after the nth clocking pulse is shown as Q_n, where Q_n stands for either of the logic levels $\underline{0}$ or $\underline{1}$. Following the next clocking pulse $(n + 1)$, the output may change, depending on the input levels S and R. With both the input terminals at logic $\underline{0}$, the clock pulse will not be able to operate either of the input NAND gates, and the output remains unchanged. A logic $\underline{1}$ at S and $\underline{0}$ at R results in the S NAND gate output dropping to $\underline{0}$ (both inputs at $\underline{1}$), while the R NAND gate output remains at $\underline{1}$ (inputs are $R = \underline{0}$, CK $= \underline{1}$). The $\underline{0}$ input to the Q NAND gate ensures that its output rises to $\underline{1}$, whatever its previous value. Both inputs to the $\bar{Q}$ NAND gate are now at $\underline{1}$, which brings $\bar{Q}$ down to $\underline{0}$, and this, being fed back into the Q NAND gate, ensures that it remains at a stable logic $\underline{1}$ output state. Thus, a logic $\underline{1}$ at S (with its complement, $\underline{0}$, at R) sets output Q to $\underline{1}$. (It is understood that $\bar{Q}$ will always be the complement of Q, in this case $\underline{0}$.)

Similar reasoning shows that a logic $\underline{1}$ at the R input (with S at $\underline{0}$) resets Q to $\underline{0}$ (and $\bar{Q}$ to $\underline{1}$).

If both inputs are allowed to go to logic level $\underline{1}$, both Q and $\bar{Q}$ will attempt to rise to $\underline{1}$ during a clocking pulse and the resulting output will be indeterminate. This condition is easily avoided by methods described later.

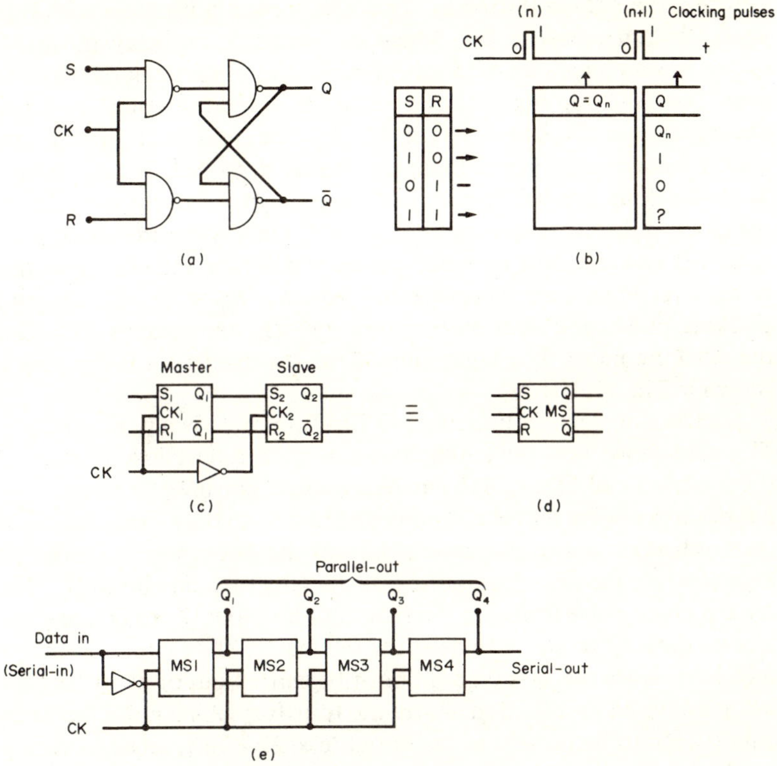

Clock pulse	Data in	Q_1	Q_2	Q_3	Q_4
1	1 →	1	0	0	0
2	1 →	1	1	0	0
3	0 →	0	1	1	0
4	1 →	1	0	1	1

(f)

FIG. 12.11. (a) The S–R flip-flop; (b) its truth table; (c) the master-slave S–R flip-flop; (d) its logic symbol; (e) basic MS shift register utilizing static logic; (f) table showing logic changes with clocking pulses.

When data is being shifted from one flip-flop to another it is important that data already in a flip-flop be allowed time to be shifted out before fresh data is shifted in. This can present a problem with integrated circuit construction, where the propagation delay through a circuit is very small and changes in input and output can overlap. To avoid this, S–R flip-flops may be connected in what is termed a Master-Slave arrangement, shown in Fig. 12.11c. The clock pulse operates the master section in the normal manner already described, but the inverter in the clocking line to the slave section prevents this operating during the clock pulse. It is said to be inhibited. Only when the clock pulse turns off can the slave operate, during which time the master section is inactive. Thus, data is transferred from S_1–R_1 to Q_1–$\bar{Q}_1$ during a clocking pulse, and then on to output Q_2–$\bar{Q}_2$ immediately following the clocking pulse. The logic symbol for the master-slave flip-flop is shown in Fig. 12.11d.

A basic shift register utilizing MS flip-flops is shown in Fig. 12.11e. This also shows how the ambiguous condition referred to in row 4 of the truth table of Fig. 12.11b can be avoided. The inverter to the R input ensures always that R is complementing S, so only rows 2 and 3 of the truth table apply. Assuming that all the Q outputs are clear (at logic level $\underline{0}$), the first clock pulse transfers the first data bit to Q_1. The second clock pulse transfers the first data bit from Q_1 to Q_2, and the second data bit to Q_1. The need for the master-slave circuit is clearly seen here, since the first data bit must be shifted out of Q_1 before the second data bit is shifted in. After the fourth clocking pulse the input data is seen to be present at the Q outputs. In effect, serial-in data is available as parallel-out data. Figure 12.11f shows a specific example for data transfer over the first four clock pulses. Data may be obtained in serial-out form at the Q–$\bar{Q}$ outputs of the last flip-flop over the next three clocking pulses.

Figure 12.12a shows how *clear* (CLR) and preset (PRE) inputs may be applied to an S–R flip-flop. The CLR and PRE inputs are normally at logic level $\underline{1}$, and must be maintained at this level during clock pulses. They may be operated at any time between clock pulses, but not together. To clear the flip-flop, the CLR input is pulsed to logic level $\underline{0}$. This raises $\bar{Q}$ to logic level $\underline{1}$ whatever the other two inputs are. This in turn ensures that all three inputs to the Q NAND gate are at logic level 1 (since both PRE, and the output of the S NAND gate are already at 1), and therefore Q drops to 0. This, being fed back to the input of the $\bar{Q}$ NAND gate ensures that the state $Q = 0$, $\bar{Q} = 1$ is stable even when the CLR 0 is removed.

To preset the Q output to 1, the PRE input is pulsed to 0. This raises Q to 1, which in turn brings $\bar{Q}$ to 0 (since all the inputs to the $\bar{Q}$ NAND gate are now at 1). The $\bar{Q} = 0$ being fed back to the input of the Q NAND gate ensures that the state $Q = 1$, $\bar{Q} = 0$ remains stable even when the PRE 0 is removed.

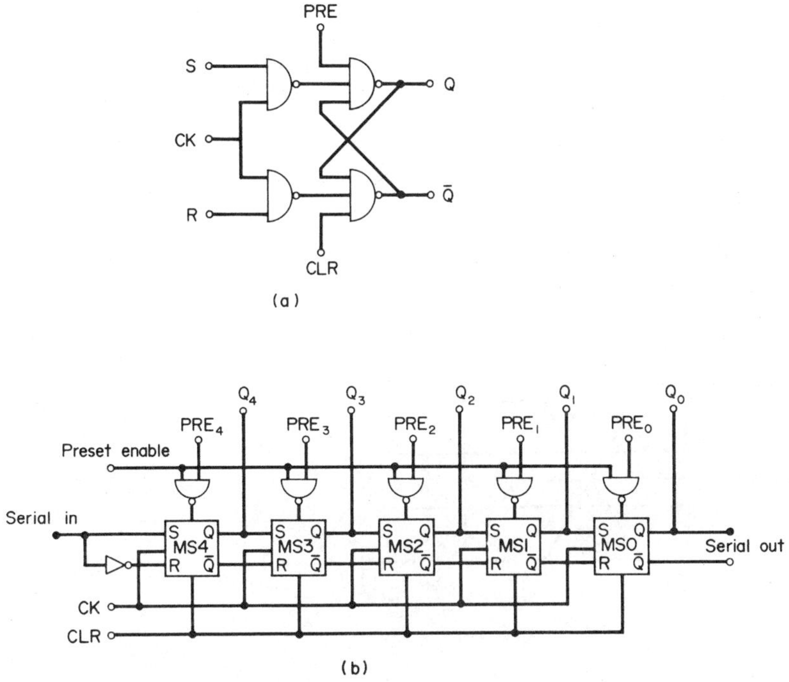

FIG. 12.12. (a) The S–R flip-flop with Preset and Clear inputs; (b) logic circuit for the TI 7496 shift register utilizing (a) type units.

Obviously CLR and PRE must not be pulsed to 0 together otherwise both Q and $\bar{Q}$ will attempt to rise to 1 and an indeterminate output would result.

In the MS flip-flop, the CLR and PRE inputs are taken to the master section. These inputs are termed *direct*, or *asynchronous*, since they are not synchronized to the clock frequency, and may be operated at any time (but not together) between clock pulses.

Figure 12.12b shows the logic diagram for a commercially available 5-bit shift register (Texas Instruments TI 7496). This utilises bipolar

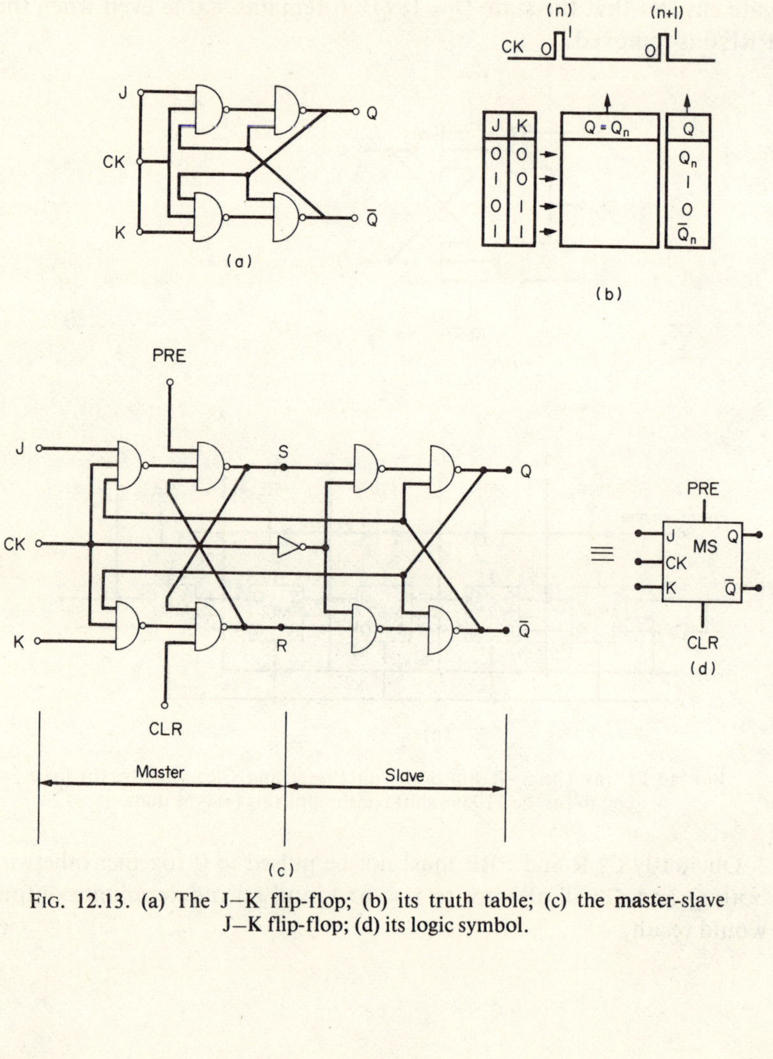

FIG. 12.13. (a) The J–K flip-flop; (b) its truth table; (c) the master-slave J–K flip-flop; (d) its logic symbol.

elements, is constructed on a single chip, and is available in a 16-pin package, and is a good example of what is termed medium scale integration (MSI). The operation is similar to that already described for Fig. 12.11e, but in addition, the NAND gates at the preset inputs enables data to be read in, in parallel. To do this the data presented to the PRE terminals, the least significant bit going to PRE_0. The register is cleared, then the PRE enable pulse is applied, at a logic level $\underline{1}$. Those NAND gates for which the data is also a $\underline{1}$, will bring the corresponding PRE inputs on the MS flip-flops to $\underline{0}$, and therefore the Q outputs will be preset at $\underline{1}$ in the manner already described. The NAND gates for which the data-in are $\underline{0}$ will not be activated and therefore the Q outputs of the corresponding MS flip-flops remain at $\underline{0}$. Thus, the parallel-in data at the PRE_0 to PRE_4 inputs appear at the Q_0 to Q_4 outputs of the register. Once in the register, the data can be shifted in serial fashion if required.

In certain counting operations which utilize shift registers it is not possible always to have the R input as the complement of the S input. To avoid the ambiguity which arises in the truth table of Fig. 12.11b, the Q and $\bar{Q}$ levels may be fed back to the input gates so that these will always have one pair of complementary inputs and therefore always complementary outputs. The basic arrangement is known as a J–K flip-flop, and is shown in Fig. 12.13a. Now with both inputs J and K at logic level $\underline{1}$ during a clock pulse, the complementary inputs Q and $\bar{Q}$ to the first pair of NAND gates result in flip-flop switching of these. In turn, the output from the first pair result in flip-flop switching of the second pair of NAND gates, (assuming Q and $\bar{Q}$ remain constant over the switching period), with the result that Q and $\bar{Q}$ finally change states. Comparing the truth table of Fig. 12.13b with that of Fig. 12.11b, and as before, denoting a Q output of $\underline{1}$ or $\underline{0}$ in general by Q_n, it is seen that the output switches to $\bar{Q}_n$ when J and K are both at logic level $\underline{1}$.

It is assumed here that Q and Q remain constant over a clocking pulse, the change only taking effect after the clocking pulse had passed. In integrated circuit construction, propagation delay through circuits can be less than a clock pulse width, with the result that changes in Q and $\bar{Q}$ are fed back to the input over the latter part of a clocking pulse. This leads to a "race around" condition in which the changes in the feedback "race" with the off-switching of the clock pulse, the final state of the output being indeterminate. To avoid this, a master-slave arrangement is used as shown in Fig. 12.13c, in which the feedback is fed from the slave section back to the input of the master section. The inverter in the clocking line to the slave section inhibits the slave during the clocking pulse, as already described in connection with Fig. 12.11c, and therefore the feedback from the slave cannot change until after the clock pulse. In this way, racing is eliminated.

The shift registers described so far utilize static logic, and are available in both bipolar and MOS technology. Dynamic logic, introduced in § 12.4 is also used in shift registers, however, for this application it appears to be available only in MOS technology. Here, it relies on the very long time-constants associated with the gate-substrate capacitance of a MOST (see § 6.12), for temporary storage of the binary data.

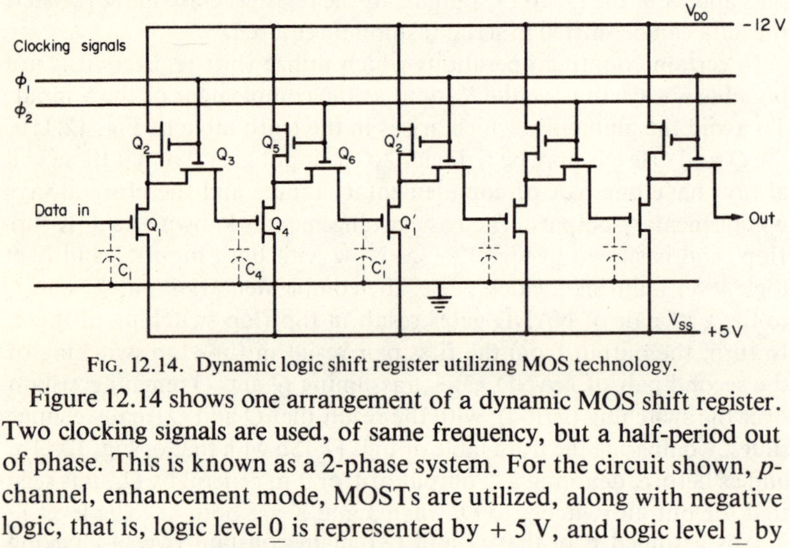

FIG. 12.14. Dynamic logic shift register utilizing MOS technology.

Figure 12.14 shows one arrangement of a dynamic MOS shift register. Two clocking signals are used, of same frequency, but a half-period out of phase. This is known as a 2-phase system. For the circuit shown, *p*-channel, enhancement mode, MOSTs are utilized, along with negative logic, that is, logic level 0 is represented by + 5 V, and logic level 1 by — 12 V. Logic level 0 switches the transistors off, logic level 1, on.

Transistors Q_1 and Q_2 form an inverter. Clock pulses ϕ_1 switch on the load transistor Q_2, and simultaneously, the transfer gate Q_3, which allows the output of the inverter to be transferred to C_4, the input capacity of Q_4. During a ϕ_1 clocking pulse, the data bit stored in C_1 (the voltage across C_1 controlling the inverter driver Q_1) is inverted by the Q_1, Q_2 inverter and transferred to C_4. When clock pulses ϕ_1 are off, clock pulses ϕ_2 are on, resulting in the load Q_5, of inverter Q_4, Q_5, and the transfer gate Q_6 being switched on simultaneously. The inverted data bit on C_4 is re-inverted by inverter Q_4, Q_5 and passed on to C'_1. Thus the logic level at C_1 is regenerated at C'_1 but delayed by the time between the ϕ_1, ϕ_2 ON pulses. The operation is repeated over each succeeding pair of clock pulses ϕ_1, ϕ_2, and in this way, data is shifted through the register.

The operation is seen to depend on charge being held on the input capacitance of each inverter. During each transfer, the capacitance is "refreshed", that is, for a logic 1 on the preceding inverter, C is connected through the transfer gate, and the inverter driver, to the + 5 V line, and for logic 0, through the transfer gate and the inverter load to the —12 V line. It is important that the refresh cycle occur frequently enough to nullify leakage from the input capacitance, and in practice this means a minimum clocking frequency of about 5 kHz.

The charge coupled device (CCD) described in § 6.13 is naturally suited to shift register applications. Figure 12.15a shows a basic 3-phase shift register, where for simplicity, surface mode operation is assumed. When the input gate is on, a conducting channel is formed between source and the potential well under the first clocked gate, which forms when clock pulse ϕ_1 is at its most positive. A charge packet proportional to the potential of the source, and therefore to the input data voltage, transfers along the conducting channel to the potential well. Switching the input gate off isolates the charge packet from the source, and it can then be transferred along the CCD in the manner described in § 6.13. The input gate is turned on by pulsing it to a voltage equal to, or greater than, the maximum clock potential, and off, by returning it to a potential equal to the lowest clock potential. Apart from these constraints, the input gate is independent of the clocking pulses.

At the output, the output gate is held at a fixed positive voltage intermediate between the gate threshold and the sense diffusion potential, and it functions as an electrostatic shield to reduce pick-up of the clocking pulses at the output. The output sense diffusion is reset every clocking period to a reference potential, by pulsing on the reset gate, as this creates a conducting channel between the reset and sense diffusions. Arrival of charge packets at the sense diffusion result in a

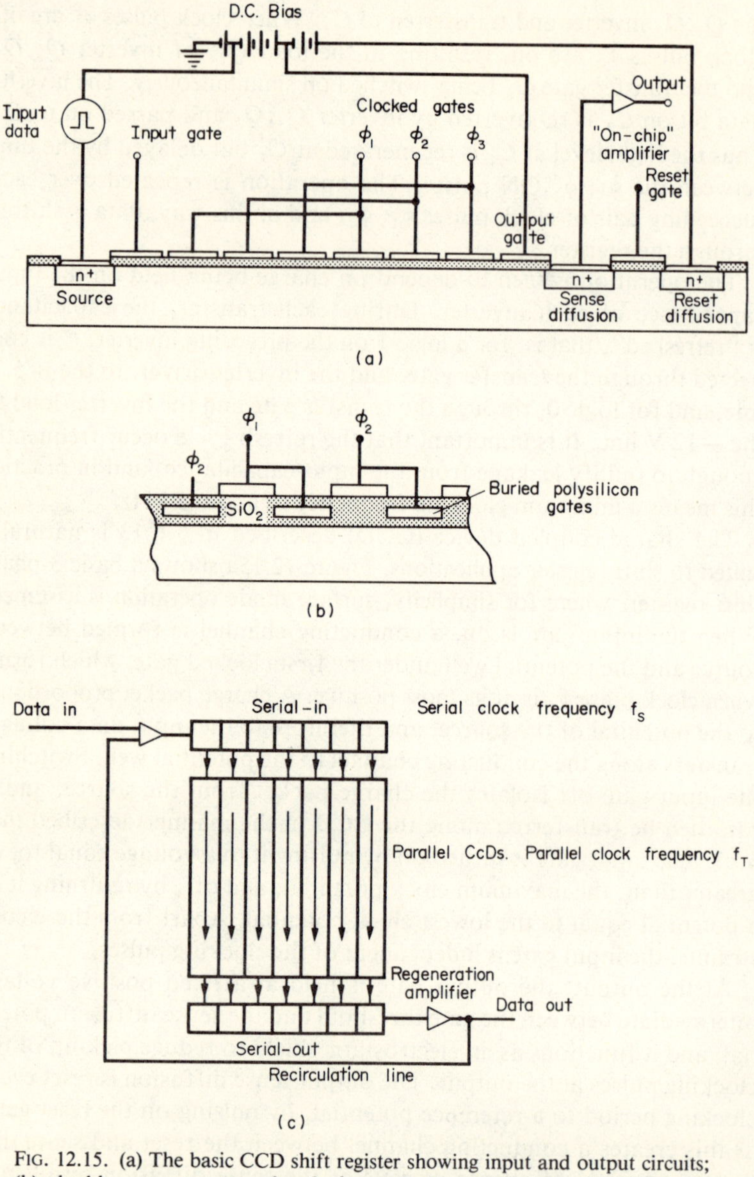

FIG. 12.15. (a) The basic CCD shift register showing input and output circuits; (b) double gate structure for 2-phase clocking; (c) Series-Parallel-Series memory organization utilizing CCD shift registers.

proportional change in potential, which is fed to an "on-chip" MOST amplifier from which the output is taken.

Two-phase clocking schemes are also used with CCDs, Fig. 12.15b showing one possible gate arrangement for this. The asymmetry in the gate structure produces asymmetry in the potential well under the gates so that charge will always transfer from left to right along the CCD. With the three-phase system, charge transfer can be reversed by suitably re-phasing the clocks, but this is not possible with the two-phase system.

Memory organization may take on one of a number of forms. The series-parallel-series arrangement shown in Fig. 12.15c is one widely used method. Here, data is transferred into a CCD in serial-in form, and this is transferred to the parallel CCDs when the serial register is filled. The transfer takes place over one clocking period of the serial CCD so the input flow is not interrupted. Two separate clocking frequencies are used f_s for the input and output serial registers, and f_p for the parallel registers. The two frequencies must be related by $f_s = K.f_p$ where K is the number of bits in the serial register.

The data appears, from the outside, to move through the register at the faster clocking frequency of the serial register, while within the register, it moves at the lower parallel clocking frequency. The power dissipation is proportional to clocking frequency, and in this way, power dissipation is reduced, but at the expense of a more complex structure compared to a straight CCD. Another advantage of the SPS arrangement compared to a straight CCD is the comparatively high bit capacity-to-transfer ratio. For a basic serial-in serial-out CCD, each data bit must transfer through each register cell. With the SPS arrangement, each data bit only transfers through the equivalent of one series CCD (the combined transfer through the input and output serial CCDs), and one parallel CCD, while the storage capacity is approximately equal to the product of the number of bits in the serial-in register times the number of bits in one parallel register.

Appendix

Commonly Used Abbreviations

Al_2O_3	Aluminium Oxide
AOI	AND–OR–Invert
BBD	Bucket Brigade Device
BCCD	Bulk (or Buried) CCD
CCD	Charge Coupled Device
CCRAM	Charge Coupled RAM
CDI	Collector Diffusion Isolation
CID	Charge Injection Device
CMOS	Complementary MOS
CTD	Charge Transfer Device
DCTL	Direct Coupled Transistor Logic
DDC	Double Dielectric Charge-storage
DIP	Dual In-line Package
D–MOS	Double-diffused MOS
DTL	Diode Transistor Logic
EAPROM	Electrically Alterable PROM
ECL	Emitter Coupled Logic
EPROM	Erasable PROM
FAMOS	Floating-gate Avalanche MOS
FET	Field Effect Transistor
GaAs	Gallium Arsenide
Ge	Germanium
HTL	High Threshold Logic
I^2L	Current Injection Logic
LID	Leadless Inverted Device
LSI	Large Scale Integration
MESFET	MEtal Semiconductor FET
MIS	Metal Insulator Semiconductor
MNOS	Metal Nitride Oxide Semiconductor
MOS	Metal Oxide Semiconductor
MOST	MOS Transistor
MSI	Medium Scale Integration
MTL	Merged Transistor Logic
n	Negative charge carrier
*n*MNOS	*n*-type MNOS
NVSM	Non Volatile Semiconductor Memory
p	Positive charge carrier
PCCD	Peristaltic CCD
*p*MNOS	*p*-type MNOS
PROM	Programmable ROM
QUIP	Quad In-line Package
RAM	Random Access Memory
ROM	Read Only Memory
RTL	Resistor Transistor Logic
SAD	Silicon Avalanche Diode
SBD	Schottky Barrier Diode
SCCD	Surface CCD
SCTL	Schottky Coupled Transistor Logic
Si	Silicon
Si_3N_4	Silicon Nitride
SiO	Silicon monoxide
SiO_2	Silicon dioxide
SOS	Silicon On Sapphire
STL	Schottky Transistor Logic
TED	Transferred Electron Device
TTL	Transistor Transistor Logic
V-MOS	Vee-notched MOS

Index